The 1998–1999 Traveler's Companions
ARGENTINA • AUSTRALIA • BALI • CALIFORNIA • CANADA • CHINA • COSTA RICA • CUBA •
EASTERN CANADA • ECUADOR • FLORIDA • HAWAII • HONG KONG • INDIA • INDONESIA • JAPAN •
KENYA • MALAYSIA & SINGAPORE • MEDITERRANEAN FRANCE • MEXICO • NEPAL • NEW ENGLAND •
NEW ZEALAND • PERU • PHILIPPINES • PORTUGAL • RUSSIA • SPAIN • THAILAND • TURKEY •
VENEZUELA • VIETNAM, LAOS AND CAMBODIA • WESTERN CANADA

Traveler's VIETNAM, LAOS & CAMBODIA Companion
First Published 1999 in the United Kingdom by
Kümmerly+Frey AG,
Alpenstrasse 58, CH 3052 Zollikofen, Switzerland
in association with
World Leisure Marketing Ltd
Unit 11, Newmarket Court, Newmarket Drive, Derby, DE24 8NW, England

Website: http://www.map-world.co.uk

ISBN: 1-8400-6069-7

© 1999 Kümmerly+Frey AG, Switzerland

Created, edited and produced by
Allan Amsel Publishing,
53, rue Beaudouin, 27700 Les Andelys, France.
E-mail: Allan.Amsel@wanadoo.fr
Editor in Chief: Allan Amsel
Editor: Anne Trager
Original design concept: Hon Bing-wah
Picture editor and designer: David Henry

Printed by Samhwa Printing Co. Ltd., Seoul, South Korea

TRAVELER'S
VIETNAM
LAOS & CAMBODIA
COMPANION

by Derek Maitland and Jill Gocher
photographed by Alain Evrard and Jill Gocher

Kümmerly+Frey

Contents

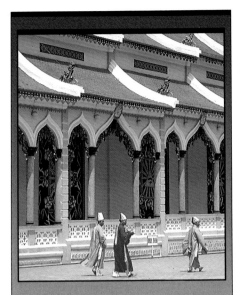

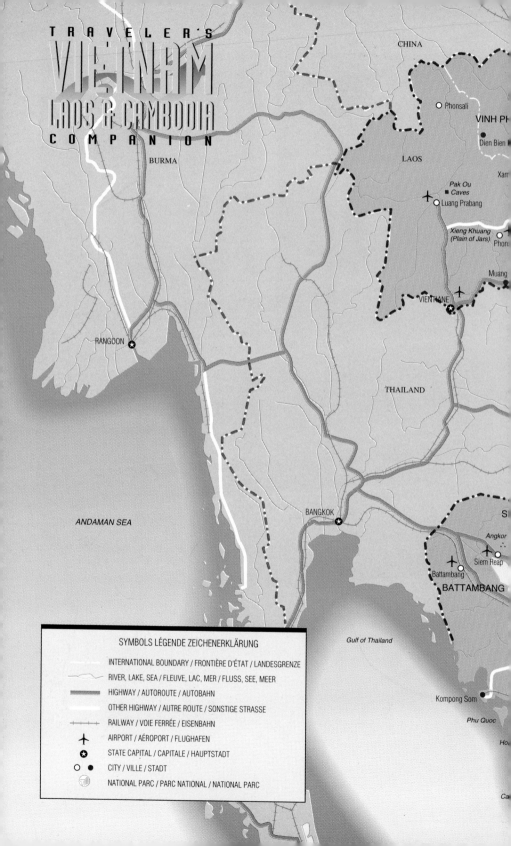

TRAVELER'S
VIETNAM
LAOS & CAMBODIA
COMPANION

CHINA

Phonsali

VINH PH

Dien Bien

LAOS

Xam

BURMA

Pak Ou
Caves

Luang Prabang

Xieng Khuang
(Plain of Jars)

Phon

Muang

VIENTIANE

THAILAND

RANGOON

ANDAMAN SEA

BANGKOK

S

Angkor

Siem Reap

Battambang

BATTAMBANG

Gulf of Thailand

Kompong Som

Phu Quoc

Ho

Ca

SYMBOLS LÉGENDE ZEICHENERKLÄRUNG

INTERNATIONAL BOUNDARY / FRONTIÈRE D'ÉTAT / LANDESGRENZE

RIVER, LAKE, SEA / FLEUVE, LAC, MER / FLUSS, SEE, MEER

HIGHWAY / AUTOROUTE / AUTOBAHN

OTHER HIGHWAY / AUTRE ROUTE / SONSTIGE STRASSE

RAILWAY / VOIE FERRÉE / EISENBAHN

AIRPORT / AÉROPORT / FLUGHAFEN

STATE CAPITAL / CAPITALE / HAUPTSTADT

CITY / VILLE / STADT

NATIONAL PARC / PARC NATIONAL / NATIONAL PARC

TOP SPOTS

Visit a Wat

SERENE AND PEACEFUL BUDDHIST COMPOUNDS, WATS (OR VATS) ARE FOUND THROUGHOUT LAOS AND CAMBODIA, and provide ideal settings to sit in the dappled shade of ancient trees and chat with saffron-clothed monks. One of the top choices on any tour of Indochina, these complexes of temples and monks' residences form a basic cornerstone of the strong Buddhist communities of these two countries. An exception is the vast Angkor Wat monument, which is not a wat in the true sense, but rather a ruined temple city. Wats are considerably smaller than Angkor and, unlike those spectacular ruins, are living complexes, supported and patronized by their devotees. They house and educate resident monks and provide spiritual succor for the people.

The basic wat plan consists of a large, walled, compound, usually sheltered by at least one bo or bodhi tree. The compound encloses an open-sided pavilion (*sala*) for sitting and receiving visitors, and dormitories (*kuti*) to accommodate the monks. A second wall often surrounds the temple (*sim*). Here the most beautiful decorations can be found, the walls painted with frescoes depicting scenes from the classic Hindu *Ramayana* epic, known as the *Pha Lak Pha Lam* in Laos.

Within the main compound are various stupas (*chedi*) and shrines, often containing holy relics, some are said to be from Buddha himself, but they also include ashes of important monks or royalty.

Laos and Cambodia abound with wats, making it easy to develop a somewhat tiresome "see every wat" mentality, the bane of many an organized tour. A more satisfying approach begins with choosing a few outstanding wats and taking the time to leisurely allow the atmosphere touch your soul.

The centuries-old compounds in Laos's royal city of Luang Prabang are especially lavish and atmospheric, an imperative on any visitor's list. Ancient whitewashed buildings are richly decorated with carvings and precious statues. Gold-leafed figures adorn lacquered pillars, ornate frescoes cover the walls of the inner

OPPOSITE: Figures from the Lao version of the Hindu epic *Ramayana* decorate temple doors in Luang Prabang. ABOVE: Monks make an alms round at the Wat Phu festival in Champassak.

temples, and low sweeping roofs are reminiscent of Thai palace architecture. Heavy wooden doors are decorated with gilded carvings of the *Ramayana*'s romantic heroes, Rama and Sita, whose various incarnations can be recognized across Indianized Southeast Asia.

Of Luang Prabang's many beautiful wats, I recommend the Wat Xieng Thong, on Manthatourath Road, and the three wats near Villa Santi — Wat Pa Phay, Wat Sieng Mouane and Wat Choum Khoung — each particularly interesting.

There are no wats in Vietnam, however, as their temples are the typical Chinese kind, with ornate decorations at every corner. You won't see saffron-robed monks roaming the streets, or indeed anywhere at all in Vietnam — except perhaps an occasional one who has wandered in from Laos or Cambodia. The Vietnamese monks wear gray and are not highly respected, as communism has pretty well obliterated the Buddhist religion in the country, apart from where it is mixed with Taoism.

Late afternoon is a pleasant and peaceful time to visit a wat, as the daily chores are completed and the younger monks and novices can be found relaxing. They are usually eager to practice their English with visitors and learn about foreign cultures and lifestyles so different from their own. It is important to remember, even in these liberal times, that a woman should never touch a monk, or even hand anything directly to him, but rather present it to a third person (male) who can pass it across. When visiting any wat, remove your shoes before entering a temple, although they can be worn in the open compounds.

Ride the Perfume River

AMONG FARMS AND RICE FIELDS,
EVOCATIVE RUINS OF AN IMPERIAL PAST DOT
THE LANDSCAPE ALONG VIETNAM'S MISTY
"PERFUME" RIVER (HUONG GIANG). The cool,
green river offers an idyllic way to
explore this United Nations World
Heritage Site where the tombs and
pagodas of the former Nguyen
Emperors lie amid rural farmlands.

Named after a scented shrub that
grows along its banks, the Perfume River
flows through Hue, one of Vietnam's
most romantic cities and once home to
Vietnam's most powerful emperors. The
Nguyen Dynasty (1802–1945) was the
first single court to rule both north
(Annam) and south (Viet Thuong) and
thus was instrumental in uniting the
country later to become Vietnam. This
not-so-ancient dynasty was established
by Emperor Gia Long, whose projects
included the Mandarin Road which
would link Hanoi to Saigon for the first
time.

Many of the major tombs, namely the
Citadel (modeled on Beijing's Forbidden
City), were all but destroyed during the

1968 Tet Offensive, although some
restoration work has been done on the
most important buildings, with more
work planned for the future. While the
scale and grandeur of the monuments
may fail to fulfil the expectations of
some exacting visitors, the superb
settings and pastoral scenery can more
than compensate for any such
shortcomings.

Be sure to include in your visit Kai
Dinh's Tombs, with its dragon pillars and
wide staircase Minh Mang's Tomb,
overlooking a particularly lovely stretch
of the river; and the pleasure gardens
surrounding Tu Duc's Tomb. Plan a late
afternoon arrival at the Thien Mu Pagoda
to watch the sun setting across the river
in the soft evening light.

To visit the tombs, join a group tour
or rent a covered *sampan* and proceed at
a gentle pace. To do so, check with your
hotel or try at the boat depot adjacent to
the Century Riverside Hotel at 49 Le Loi

OPPOSITE TOP: Sparkling white
stupas of Wat Botum, Phnom Penh.
OPPOSITE BOTTOM: Offerings are placed
before effigies of the Buddha to pay respects.
ABOVE: A pleasure pavilion looks across
the placid waters at Tu Duc tombs, Hue.

TOP SPOTS.

Street, where boats are available for US$30 to US$40 for the day. While simple food is available at food stalls near some of the tombs, I recommend taking a picnic basket from your hotel and choosing a picturesque spot for lunch.

To relive the days of Vietnam's most powerful rulers, you can rent a decorated and kingly barge for the evening and dine on the river. Guests are served the exotic courtly dishes once served to emperors, and entertained with traditional Vietnamese folk music. Several hotels, including the Huong Giang Hotel ((54) 822 122, 51 Le Loi Street, will organize floating dinners.

Explore the Tonkinese Alps

VIETNAM'S HIGHEST PEAK, MOUNT FANSIPAN, RISES 3,142 M (10,309 FT) FROM THE RUGGED MOUNTAIN HIGHLANDS NEAR THE CHINESE BORDER ABOVE THE OLD FRENCH COLONIAL HILL STATION OF SA PA. The Hoang Liem Nature Reserve surrounds a conglomerate of mountains that includes Mount Fansipan. It takes a minimum of three days to hike up and down the peak, but it's better to allow six days. The cool climate and surrounding farmlands and hill-tribe villages lend themselves to trekking and long country walks.

While the winter months of December to March are cold and often very misty, springtime (April to May) brings plum blossom and magnificent views across the mountains. The best season for hiking is during the months of October to April, when the air is coolest and the main rainy season has passed.

Sa Pa provided one of my favorite images of Vietnam, capturing something of its spirit. One cold winter night I walked towards my hotel, the mist swirling around in thick clouds, severely limiting visibility. From one lowly eating house a dim circle of yellow light spilled out, illuminating a group of young Hmong girls huddled together in the

freezing air, chattering happily, intently watching the goings-on inside, a scene reminiscent of some medieval drama.

An old trading town in the center of a region populated with hill tribes, Sa Pa's streets fill daily — particularly on Saturday — with villagers coming to the market with their produce and, more frequently now, tribal clothing and handmade souvenirs for the growing number of tourists.

Panoramas of Mount Fansipan are best viewed from the Victoria Hotel ((20) 871 522 FAX (20) 871 539, Sa Pa's brand new French-run resort, one of a few new resorts found outside the capital cities. Guests can spend evenings around open log fires sipping French wine and dining on cheese soufflés and hearty *pot au feu*.

While tours to Sa Pa are readily available from Hanoi they are quite unnecessary. Trains depart each evening around 9 PM from Hanoi Station, arriving at Lao Cai, near the Chinese border, at around 7 AM. From the station counter you can buy a tourist bus ticket to Sa Pa for a nominal sum and embark on the headily panoramic 45-minute drive through the mountains to the village.

Along with the Victoria, the small hotels around the bazaar on the main street offer tours and rent jeeps (or battered old Russian four-wheel drive vehicles). Treks to Mount Fansipan and day trips to nearby villages can be organized easily. The Victoria looks after its guests and will organize whatever they would like to do. Guest houses hang signs at their entrances advertising their tours, all of which are similar. The Dang Trung Auberge ((20) 871 243 FAX (20) 871 282 is run by its charming French-speaking owner, Mr. Dang Trung, who takes delight in showing his magnificent rock garden to guests. His competent staff can organize treks and tours around Sa Pa, tickets to Hanoi, four-wheel drive vehicles, trips to Bac Ha, and any other adventure one might like to partake in.

A Red Zao woman, member of one of the Sa Pa region's many minority groups.

Marvel at the Stone Jars

SCATTERED ACROSS THE ARID PLAINS OF LAOS'S XIENG KHUANG PROVINCE ARE HUNDREDS OF GIANT JARS, INTRIGUING VISITORS WITH THEIR UNEXPLAINED ORIGINS. These enigmatic antique stone jars have captured the imagination of all who visit them on the stark alluvial and bomb-scarred plains of northeastern Laos. The Plain of Jars is a "marvel... a mystery... a site unique in all the world, due to its archeological value, its cultural specificity, the unknowns that it hides within," according to Mr. Frederico Mayor, leader of a February 1998 UNESCO delegation which allocated US$50,000 for the Plain's rehabilitation.

The jars stimulate numerous wild and increasingly speculative theories, and you'll find everyone has a favorite. With the largest jar reaching over three and a half meters (12 ft) in height, and the smallest roughly half a meter (two feet) high, one can only wonder.

There are no stone quarries in any near vicinity, so where did the stone come from? Are they carved from stone

at all, or are they created from some long forgotten recipe for porous cement? Why did they survive the countless B-52 raids over what was probably the worst hit area in the whole of Laos during the Vietnam War? Are they the remnants of some ancestor-worshipping cult, sarcophagi for the rulers of an ancient tribe (It is said that in the 1930s French archeologists found human bones beside some of the jars)? Were they used to store water or was there once a tribe of monumental drinkers? Or (of course), was it a band of extraterrestrials who left these as the only evidence of their visit to earth? Adding to the mystery, on the Indonesian island of Sulawesi there are similar jars also scattered over a remote and almost inaccessible plain.

To visit the Plain of Jars, most visitors fly to Phonsavan from Vientiane en route to Luang Prabang, or vice versa, staying a night in a reasonable, although not luxurious, hotel, before heading to their next destination the following day. Those not on a tour will find the visit easy to arrange upon arrival at Phonsavan Airport, where guides and drivers are eager to assist with all arrangements for a small fee.

Ride the Reunification Express

TRAIN AFICIONADOS CAN'T AFFORD TO MISS VIETNAM'S THONG NHAT, OR REUNIFICATION EXPRESS, whose 2,012-km (1,250-mile) journey takes you from Saigon to Hanoi or back again in an epic 48 hours on the S7, or a speedy 36 hours on the S3/CM5. Whichever speed you travel, the ride will prove memorable.

The rather inaptly-named Express recommenced operations on December 31, 1976. Originally built by the French in the late 1930s, it was bombed incessantly during the war, to nearly total destruction. Relaunching the railway was the culmination of months of restoration, involving the repair of 158 stations, over 1,330 bridges, nearly 30 tunnels and 1,370 shunts. This single track was one of the first projects undertaken after the 1975 reunification.

While the train runs almost the length of the country from north to south, it is possible to take just one or two legs of the journey or break the trip along the way. Hanoi to Hue is a pleasant overnight ride which arrives around midday. Hue to Da

Nang provides a good five-hour ride through luscious countryside.

Although service has improved over the years, riding in the train is still an adventure, a chance to see the country in comfort, yet still experience the everyday life of the Vietnamese. Stopping at stations is always a noisy adventure, as hawkers crowd to the barred train windows selling freshly boiled eggs, bottles of drinks, tea and coffee, exotic-looking fruit, bowls of steaming noodles, and mysterious packets of food. As night falls, people in the cheaper, hard, seats start hanging up their hammocks, until the carriages and passage ways are festooned with a maze of swinging bodies, comfortably ensconced in folds of dark green spun polyester. Several concessions have been made for tourists — namely the price is more than double what the Vietnamese pay; but in Vietnam, this is the norm.

The compartments vary from hard seat and soft seat to hard sleeper, where you get a wooden rack in a cabin of six,

OPPOSITE: The mystery of Laos' giant stone jars remains unexplained. ABOVE: Rudimentary food stalls sell delicious food in northern Vietnam.

to four-berth soft sleeper, which is the most comfortable. If you are not traveling in a party of four, trust to luck and the good will of the booking staff for your travel companions. I have experienced some delightful trips with Vietnamese companions, although the tendency now is to group foreigners together.

The trip includes seeing a large amount of the country in comfort. Those who have the foresight to bring a picnic or food hamper will find it even more rewarding. The best time to go is during the cooler winter months when you can snuggle in under the warm blankets provided. Air conditioning is not an option. Highlights include the panoramic Hai Van Pass, or Cloud Mountain Pass, where an extra engine is added to pull the train up the long and winding track before descending again to the coastal plain.

Sample a Snake

YOU CAN DELVE INTO THE EXOTIC SIDE OF SOUTHEAST ASIAN CULTURE BY SAMPLING A PLATE OF SNAKE WASHED DOWN WITH A GLASS OF SNAKE WINE. A popular culinary treat for the Vietnamese, snake snacks are available in specialty restaurants across the country. Snake eating ranks among macho activities especially favored by Vietnamese men, who consider the reptile meat to be a "heaty" food on par with dog and monkey as far as aphrodisiacs go. The liver is said to be the most potent. Also taken as a restorative, blood from the freshly-killed snake is mixed with brandy and sipped for renewed strength. In one country town I visited, the local rice restaurant had large jars of pickled snakes sitting up on the counter, their beady eyes staring at potential customers. This snake-soaked wine is also considered to be a health tonic.

About five kilometers (three miles) from Hanoi is Le Mat village in Gia Lam, known for its ten or so snake restaurants that will dish up a feast of serpents. Try Quoc Trieu ((73) 827 2988,

where a cobra-filled snake pit offers plenty of choice.

For a serious journey into the snake world, try the Dong Tam Snake Farm on the outskirts of the Delta town of My Tho, 72 km (45 miles) from Saigon, on the road to Vinh Long. The small museum holds 40 varieties of snakes stored in jars of alcohol, while the nearby Military Zone Snake Camp, set up as a medical research center, has live snakes, particularly cobras and pythons, to inspect. Afterwards, ask your driver to take you to the nearby Dong Tam Restaurant ((73) 873 494 to sample all manner of snake-based specialties.

For a less serious venture, a group of tourist restaurants in the delta town of Can Tho, stretched along the main drag of Hai Ba Trung Street facing the river, serve grilled and fried snake on the menu along with more mainstream dishes of fried noodles. The squeamish can feast on delicious seafood treats of crab, prawns and fresh fish. For the curious, snake tastes just like chewy chicken, which is not too bad if you can stop the visions of slithery reptiles while you chew.

Ride the Mekong

THE MIGHTY MEKONG, WHICH FORMS PART OF THE BORDER BETWEEN OF THAILAND AND LAOS BEFORE ENTERING CAMBODIA AND FINALLY THE VIETNAM DELTA, IS NAVIGABLE FOR MUCH OF ITS 4,200-KM (2,600-MILE) JOURNEY TO THE SEA. One doesn't need to be an intrepid explorer to enjoy what the Mekong has to offer, and even a short ride can evoke a feeling for this powerful river. The highway of the past, river traffic is currently diminishing, as it gives way to the far less romantic road travel. Perhaps tourism will instigate its revival, as Laos strives to make river tourism a major offering in the future. In addition to the Mekong, other rivers such as the Sekong in the south and the Nam Tha and Nam Ou in the north offer memorable adventures for those seeking interesting alternatives.

Trips, tours, and privately chartered boats abound in different parts of Laos and can be easily integrated into an organized tour of the country, or added on once in the vicinity. Half the visitors to

northern Laos now enter through the Thai border at Ban Huay Sai before embarking on the two-day Mekong trip downstream to Luang Prabang. Appreciated by backpackers, this trip is also part of an upmarket tour run by Diethelm, Intrepid Travel, and a host of other tour operators (see TAKING A TOUR, page 67).

An easy day trip runs from Luang Prabang north to Pak Ou Buddha Caves, a ride of an hour or two which stops at a Hmong village along the way. Boats can be rented at the boat jetty or through an organized tour easily arranged at your hotel tour desk.

Visitors with a taste for basic travel can take a three-day downstream, or four-day upstream, river cargo boat from Vientiane to Luang Prabang. Although not a tour option at this stage, the trip can be arranged by making inquiries at the boat jetty in either Vientiane or Luang Prabang. Costs are minimal, as are the facilities.

Mekong Delta: although a little unappetizing, snake wine is believed to impart strength and vigor.

The memorable three-hour ride from Pakse to Champassak in southern Laos, the gateway to the pre-Angkor Wat Phu, is by far the most enjoyable way to reach the ancient complex, which is then only a four-kilometer (two-and-a-half-mile) cyclo or taxi ride away. Visitors can jump in with the locals on a public river ferry for a few kip, or pay more and charter a boat for a few dollars. For those who prefer an extra feeling of security, tour operators can organize trips for groups. Once you are in Pakse, Sodetour ((31) 212 122 (whose office is close to the river confluence on the Don River side) can organize the trip to Champassak and Wat Phu or even to the Mekong Islands. Alternatively, it is possible to contact their Vientiane office ((21) 213 478 FAX (21) 216 313 at 114 Quai Fa-Ngum.

In Cambodia you can book a half-day Mekong trip through farmlands from the tour desk at the Sofitel Cambodiana Hotel ((23) 426 288 FAX (23) 426 392 at 313 Sisowath Quay in Phnom Penh.

Witness a Cultural Revival

ONE OF THE SPECIAL TREATS IN PHNOM PENH IS TO WITNESS THE CREAM OF CAMBODIAN YOUTH STUDYING THE SKILLS OF THEIR ANCESTORS, learning to embody the heavenly nymphs, or *apsaras*, of Hindu mythology as they are portrayed in classical Khmer dance, part of a budding cultural revival you can witness at the School of Fine Arts on Rue des Petites Fleurs (Street 70).

In Year Zero (1975), when the Khmer Rouge tried to radically restructure Cambodia into a peasant-dominated agrarian economy, almost all of Cambodia's intelligentsia, artists, craftsmen, and talented purveyors of the arts were brutally killed, leaving a large gap in the country's cultural wealth. Fortunately a few survived. Perhaps they fled the country, perhaps they kept a very low profile. In any case, Cambodia is currently experiencing a revival of the arts. Once again beautiful young girls are learning the disciplined and graceful

moves of Khmer classical dance which echo the celestial nymphs depicted in carvings in the many galleries and temples of Angkor Wat. It is a rare privilege to watch the heartening sight of a cultural revival.

This is no performance for the benefit of tourists, but a window into their culture — a culture that was once anathema to the Khmer Rouge who did everything in their power to destroy it. Classes consist of about forty students whose ages range from five or six to sixteen. While boys also learn to dance, and play important roles, they are fewer and less elegant and refined than the girls.

To gain entry, arrive around 7:30 AM, and if you are without guide, simply request permission from the gate sentry, if one is on duty. The actual dance studio is quite informal and as long as you remember to slip off your shoes and just take a seat at one of the benches you will be set for an absorbing hour or two. Not many visitors make this worthwhile visit.

OPPOSITE: Craft and produce of all descriptions attend the bustling floating market of Cai Rang in the Mekong Delta. ABOVE: Banteay Srei is well known for its finely executed sandstone carvings like this delicate apsara.

Admire Banteay Srei

THE PINK CITADEL OF WOMEN IS ONE OF THE MORE RECENTLY OPENED TEMPLE COMPLEXES SURROUNDING ANGKOR WAT. It goes without saying that a visit to Angkor Wat is one of the imperatives of any Indochina tour, but not as many visitors are aware of the fabled pink Citadel of Women known as Banteay Srei. It had been inaccessible since the 1970s because of the Khmer Rouge threat, but with the Cambodian government's efforts to remove this threat the temple finally opened to visitors in 1997, making it a must-see on the temple tour circuit. The Shivaist Banteay Srei Temple is located 32 km (20 miles) from the main Angkor Wat complex, a drive through the rural Cambodia of rice fields and simple, stilted houses.

Dance is just one of the arts that has somehow managed to survive; in parts of the country you may be lucky enough to stumble upon a rural performance of masked theatre or a shadow puppet show which incorporates stories from the classic Indian epic, the *Ramayana*. Also, the Ministry of Culture and Fine Arts ((23) 362 647 on Croix Rouge Khmer Street (Street 180), is sometimes able to organize special dance and music performances by the students.

On Wednesdays and Saturdays, classical Khmer dance performances are held in Siem Reap at the Performance House in the Grand Hotel d'Angkor ((63) 963 888. Tickets for the buffet dinner of traditional Khmer food (7 PM) and the performance (8:15 PM) cost US$15.

Otherwise, the best place to witness classical dance is during the Ramayana Festival held each November at Angkor Wat, on the day coinciding with the full moon. Classical dancers from the surrounding countries of Indonesia, Myanmar, Laos, and Thailand are invited to participate in this celebration of dance and the arts.

The temple's intricate, finely executed, pink sandstone carvings, and its depictions of heavenly deities, both male and female, and celestial nymphs, or *apsaras*, make it one of the region's most beautiful temples. It is also one of the oldest: it was dedicated in AD 987, during the reign of King Rajendravarman. The temple is filled with the finest carved stone, every inch of the surface covered with details of trailing flower tendrils interlocked with leaves, geometric patterns, and intricately executed flutings. Scenes from the Indian classic tale of the Ramayana are illustrated in detail, one depicting the much despised and feared evil Rahwana making off, spear in hand, with the beautiful Sita, wife of the good King Rama. Stolid temple guards, clubs in hand, stand guard over the entrances.

Possibly because it is older, Banteay Srei was built on a much smaller scale than Angkor Wat. Where Angkor is a multilevel complex, Banteay Srei is a far simpler one-level concern, its main entrance lined with what were once giant stone standing lingas, many of which have since fallen into disrepair.

22

Best of all, because of its distance from Angkor only the most dedicated visitors make it here. It is best visited in the cool of the early morning, before the heat of the day, or in the late afternoon. Tours are easily arranged in Siem Reap from your hotel tour desk, or incorporated into a tour of Angkor. Check with your hotel on prevailing safety conditions.

Haggle in a Country Market

SOME OF ASIA'S BEST MARKETS CAN BE FOUND IN INDOCHINA. Where rural life is strong, markets are a major event — the country equivalent of a pub, shopping mall, coffee shop, restaurant, supermarket and disco rolled into one. This is the place where folk can catch up on local gossip, sell their wares for much-needed cash, hunt for a husband, discuss the intricacies of cattle/horse/chicken breeding, eat out, and generally enjoy a break from the hard routine of village life.

And markets are a fine place to practice your haggling skills. Bargaining is an expected part of every transaction in Asia — I even bargained down the price of an operation in an Asian hospital with

very gratifying results. Basic bargaining rule number one is: Keep it pleasant. We bargain as sport, but to the people we bargain with, it represents a livelihood. When faced with a language barrier, practice nonverbal communication. Hand signs, gestures, fingers (for counting, although the more sophisticated use calculators), grimaces and smiles can all be brought into play, and in the end it can become a highly amusing game. The market streets of Sa Pa are a very good place to put newly gleaned skills to use as traditionally dressed tribal women approach selling hand-embroidered costumes and handicrafts.

Bac Ha in northern Vietnam has one such memorable market. It is best on Sundays when the lower reaches of the small town swarm with tribal villagers known, in a loose translation, as the "Variegated" or "Flower" Hmong because of their colorful costumes.

Moving to Central Vietnam, Hue's big morning market across the Perfume River from the major hotels is a

OPPOSITE: This stone temple guard stands watch over the ruins of Banteay Srei temple. ABOVE: Luscious fresh produce is sold every day in Phnom Penh's Russian market.

Vietnamesque delight. Sampans filled with cone-hatted women converge at the stone steps leading to the market. The women make their way through the crowds bearing great baskets of vegetables, or perhaps a squealing pig or two tucked under their arms. My Tho, in Vietnam's Mekong Delta, has a huge area given over to daily commerce. Myriad varieties of rice are piled into neat mountains at each stall, and vegetables gleam with color and freshness as ducks are brought to market hanging in

bunches off bicycles, quacking wildly while the owner fights for a space. The bustling floating market in Phung Hiep, a small market town 35 km (20 miles) from Can Tho, also in the delta, is a riot of small sampans piled with fresh produce, each negotiating its way delicately through the crowded maze of craft. The morning market in Hoi An also bustles with activity and color. Sampans loaded to the gills with fresh seafood of every conceivable variety are floated in to tie at the main jetty adjacent to the market.

In northwest Laos, close to the Chinese border, is the tiny village of Muang Sing, a mecca for hundreds of tribal villagers in the district. Wearing their tribal best, villagers converge on the market each morning to sell their wares. Many come on foot, but the lucky ones come proudly bundled into a wagon attached to their communally-owned tractor.

Visitors who have no time to venture into rural areas can enjoy a similar spectacle in Laos's Luang Prabang morning produce market, where dogs are butchered along with pigs and chickens. For any market visit, remember that villages arise early, and in many rural areas the action is all over by around 9 AM.

Contemplate Indochina's Recent Past

AMIDST THE BEAUTY AND GRACIOUSNESS OF INDOCHINA, CHILLING REMINDERS OF THE RECENT PAST REMAIN, a past too near and too atrocious to ignore. Almost as if called upon to create some kind of yin-yang balance with the abundant beauty of this region, sobering remnants of a darker side bear witness to man's inhumanity to man.

In Hanoi, the severely truncated Hoa Lo Prison, or "Hanoi Hilton" as it was known to the Americans, has opened its 100-year-old doors once again, this time as a museum. Above the sturdy wooden doors are two words, "Maison Central," that once struck terror into the hearts of anyone entering it, on what was frequently a one-way journey. Although most of the original prison grounds have been demolished to make way for the high rise Hanoi Tower, enough remains of Hoa Lo Prison for visitors to experience the chilly conditions of what was once Indochina's second largest, and Vietnam's most notorious, prison.

At times Hoa Lo housed up to 2,000 prisoners. Ironically, while it was built by the French, it was later used by the Vietnamese to hold their own political prisoners. The last prisoners left Hoa Lo in 1994, and although it is now safe to enter, the grim reminders of over 100 years of pain and torture and heroism remain. One heartening prison story relates how, in 1945, more than 100 Vietnamese political prisoners escaped through two drainage gates and a muddy tunnel. One of these escapees was Do Muoi, who later became a country leader.

Another important institution is Saigon's War Crimes Museum, where giant stark black-and-white enlargements record famous instances of the inhumanity and futility of war. Amongst the exhibits are Purple Crosses and other military badges of valor returned to the Vietnamese government by United States soldiers with letters of apology to the Vietnamese people.

In Cambodia one only needs to walk the streets of Phnom Penh to see countless living reminders of war — one-armed beggars, severely truncated men, and hordes of homeless children roaming the streets, congregating outside the National Art Museum in hope of handouts. Inside the Museum, in a macabre repetition, is a similarly truncated, and headless, statuary — victims not only of the war, but also of art thieves and Khmer Rouge opportunists.

I am always surprised that such seemingly placid people as the Khmer were capable of the atrocities of their recent past. Phnom Penh's Tuol Sleng Museum, which has been described as "Auschwitz on the Mekong," graphically portrays the sickening excesses of Pol Pot's madness. This onetime high school became Security Prison 21 (S-21), a detention and interrogation center from which only a handful of detainees ever emerged alive. Inside, wall after wall of stark monochrome photos depict prisoners before and after torture, many of whom were then killed in the school compound — with clubs to save on bullets.

OPPOSITE: Detail of a finely woven handloom shawl TOP with silk accents, from Luang Prabang. Steaming bowls of *pho* noodle soup BOTTOM at the early morning market in Muang Sing. ABOVE: United States helicopter at the Museum of American War Crimes in Saigon mixes with ordnance and other detritus of war.

YOUR CHOICE

The Great Outdoors

In Indochina, the Great Outdoors and such notions as trekking or enjoying a walk in the rainforest remain little-developed tourist concepts. In a region where the majority of the population depends on a rural subsistence economy entailing daily hours of backbreaking labor, the Great Outdoors is perceived as something to be farmed rather than to be explored during leisure hours.

For the majority of Asians, the wilderness becomes attractive only after it has been tamed. In the few outdoor-related tourist areas, such as Central Vietnam's Valley of Love in Da Lat, local tourists rush to have their pictures taken against kitsch concrete statues and

memorials, with larger-than-life Disney characters, or sitting atop a miniature horse wearing a cowboy hat. On Vietnam's remote Cat Ba Island, the small township is crowded with karaoke bars and souvenir shops, and few local tourists ever make it to the national park on the other side of the island.

In addition, much of the Great Outdoors is still alive with land mines, which continue to cause dreadful casualties, especially among children and farmers. While areas are being cleared, the work is by no means completed. Laos's Xieng Khuang Province (and the Plain of Jars) and the areas around Angkor Wat in Cambodia have been sufficiently cleared to be considered safe to visit without sticking to the well-marked paths, although it is advisable to be especially wary in the more remote areas away from important tourist sites. There is no need for great fear, but take a guide and proceed with awareness.

However, that said, outdoor areas and national parks do exist and lush tropical wilderness can be visited by tours or by individuals if they have transport. While sophisticated facilities are often lacking, better-known national parks, especially in Vietnam, have at least one guest house and a few guides waiting for business.

OPPOSITE: Waterfall by the That Lo resort , Boloven plateau, southern Laos. ABOVE: The placid blue waters of Nha Trang's beaches attract vacationers from around the world.

NATIONAL PARKS

While Indochina has designated national parks, many are not yet accessible or have minimal facilities. Illegal logging continues to encroach on primary forest, and, in Vietnam, the years of war brought vast ecological devastation. Nevertheless, despite the statistics the picture is not all dismal. Flying from Vientiane to Luang Prabang in Laos one sees nothing but mile after mile of primary forest covering steep and rugged mountain ridges, with the very occasional clearing for a small tribal village.

VIETNAM

With the creation of 87 national parks, Vietnam is among the most advanced nations ecologically speaking, although percentage figures are far less optimistic. Only 29% of the country's forested land remains, and more forests are disappearing with the current land clearings for coffee plantations in the Central Highlands. Eight primeval forest areas have been classified as national parks — these are Cuc Phuong, Cat Ba Island, Ba Vi, Ba Be Lake, Bach Ma, Cat Tien, Con Dao Islands and Phu Quoc Island. Especen Travel offers tours to several of these national parks (see TAKING A TOUR, page 67).

In the far north of Vietnam, in the tribal trading town and French colonial hill station of Sa Pa, close to the Chinese border, it is possible to hike in the **Hoang Liem Nature Reserve** which contains Vietnam's highest peak, the 3,142-m (10,309-ft) Mount Fansipan (see EXPLORE THE TONKINESE ALPS, page 15 in TOP SPOTS). Guides can be found in Sa Pa or a more organized tour can be taken from Hanoi.

Also accessible from Hanoi is **Cat Ba Island** and the surrounding archipelago. Beautiful beaches and grottos mark the island coastlines, while thick old-growth forest covers much of this 570-hectare (1,400-acre) national park in Vietnam's Ha Long Bay, 80 km (50 miles) east of the mainland town of Hai Phong. Consisting of 366 islands and islets, the Cat Ba archipelago spreads across the bay.

WATERFALLS

Waterfalls are popular with everyone, and locals and foreign visitors alike enjoy visits to the many spectacular falls scattered through often remote areas in Indochina. With so many rugged and untouched mountain ranges, Laos is blessed with an abundance of fine examples, some of the most spectacular ones requiring a special trip to reach. Some originate in the deep jungle-covered mountains of the **Boloven Plateau**, falling near the Sekong River, an area toured by Diethelm (see TAKING A TOUR, page 67) on one of their southern Laos circuits. Near Paksong, you can visit the 130-m (425-ft) **Tad Phan Falls** and the 10-m (33-ft) **That Lo Falls** where the pleasant bungalows at the **That Lo Lodge (** (31) 212 725 overlook the falls.

Others, such as the **Kuang Xi Falls**, 30 km (19 miles) south of Luang Prabang, make an easy day trip by motorbike or jumbo from the That Luang Market in Luang Prabang. In Vietnam's Central Highlands close to Buon Ma Thuot is the **Dray Sap Waterfall**, 25 km (15 miles) out of town, on a good road, and the **Krong Bong Waterfall**, 45 km (28 miles) from town, past the village of Lak.

Hornbills, reptiles, and mammals including wild cats, gibbons, boar, monkeys and deer inhabit the pristine forests. Mangrove forests and freshwater swamps are home to a diverse range of water birds.

Cat Ba town consists of a huge population of boat dwellers, filling the broad horseshoe bay. The bay road is lined with towering mini hotels to accommodate an increasing influx of tourists. Waterfront cafés offer tours, motorbikes, and national park guides. The park is around 15 kilometers (five miles) from the small town; the best tour leaves early and includes a five-hour walk over seven rugged peaks to a secluded beach. A boat returns you to the village after lunch.

The island is accessible by fast boat from Hai Phong and also by a slower ferry (which takes around four hours) from the Cat Ba ferry terminal at Hai Phong's harbor. Hai Phong is accessible by train from Hanoi — the recommended way to travel.

The **Ba Be Lake** (pronounced bar bay) is a clear-water lake filled with rare fish and abundant colonies of aquatic bird species, surrounded by limestone karsts making it one of Vietnam's more attractive national parks, and covering an area of 23,340 hectares (577,670 acres).

OPPOSITE: Small chalets sit by a sparkling river and waterfall at the That Lo Resort in southern Laos. ABOVE: The crumbling ruins of Cham towers stand proudly by the roadside south of Nha Trang in central Vietnam.

Trails lead through the park, and to the villages of several minority groups. Ba Be is a full day's trip from Hanoi, so at least three days are required for the visit. It can be incorporated into a tour of the north or as a trip on its own. Especen Travel Agency and Saigon Tourist (see TAKING A TOUR, page 67) handle tours, and there is simple accommodation within the park confines. Independent arrivals will find guides and transport awaiting them.

The 25,000-hectare (61,770-acre) **Cuc Phuong National Park** lies 140 km (90 miles) southwest of Hanoi and is one of the few remaining stands of primeval tropical forest. The park is home to 2,000 different plant species, including rare trees, and to 64 animal species, including yellow monkeys and flying lizards, although the chances of seeing these rare creatures are fairly slim. Accommodation is available within the park.

In the south, the **Cat Tien National Reserve** lies about 140 km (90 miles) north of Saigon, adjoining the provinces of Dong Nai, Song Be and Lam Dong in the Central Vietnam Highlands. The reserve covers an area of some 10,000 hectares (24,700 acres), the natural habitat of rare Asian species including the almost extinct Asian rhinoceros, along with pythons, crocodiles, elephants, and flying squirrels. The reserve is accessed from the main Da Lat road north of Saigon, and simple accommodation and guides are available within the park confines.

In southern Vietnam's Dong Thap Province, the 7,500-hectare (18,500-acre) **Tram Chim Reserve** in the Mekong Delta is the domain of nearly 150 species of birds, including 15 rare species. But the main attraction is the migrating cranes that come to the sanctuary around December and stay until April, the season dry enough for the one-and-a-half-meter (five-foot)-tall sarus cranes to walk around the swamp and dig for food. Other birds include black drongoes, purple and gray herons, black-shouldered kites, purple swamp hens, cormorants, kingfishers, and great egrets (see TAKING A TOUR, page 67).

Also located within the vast Mekong Delta is the **Tam Nong Reserve**. Covering around 750 hectare (1,850 acres), the reserve is home to almost 150 species of birds, including flamingos and thousands of storks during the wet season.

CAMBODIA

In 1993, the king of Cambodia, His Majesty Norodom Sihanouk, issued "The Creation and Designation of Protected Areas" decree, which established a system of 23 protected areas covering 3.4 million hectares (over 8.4 million acres) of the country, including national parks and wildlife sanctuaries. As tourism development has been somewhat disturbed in Cambodia for some time, the parks are nowhere near as developed as

those of Vietnam. Perhaps Cambodia's most accessible park is the **Kirirom National Park** in the southern province of Kompong Speu. Within the park is a visitor's center; Cambodia's highest mountain (the 1,771-m (5,810-ft) Mount Oral); and a vast collection of wildlife including elephants, monkeys, snakes, barking deer and the occasional leopard and tiger.

LAOS

Wishing "to avoid the negative environmental impact which often results from tourism," the Lao Government has established good conservation policies for their abundant wilderness areas, and in 1993 it established legal protection for 17 areas, mostly in the south. Known as NBCAs, or National Biodiversity Conservation Areas, they cover a total of 24,600 sq km (nearly 9,500 sq miles) — around one tenth of the country's total area. Although strictly speaking they are not national parks, complete conservation areas are included within this category. Tracts of pristine rain forest are inhabited by wild species which include the leopard cat, Javan mongoose, Malaysian sun bear, Saola ox, gibbon, langur, goat-antelope, and a newly discovered rare deer antelope, along with a particular species of ox (Vu Quang) found in the east.

Wild horses are herded along a road on the Boloven plateau in South Laos.

Forests still cover large parts of the country, with an estimated 50% being primary monsoon forest. Whether they will withstand the impact of avaricious Thai logging concerns and developers remains to be seen.

Of the wild areas, the place that is being held up as a "model ecotourism resort" is the **That Lo Resort** ((31) 212 725 in Saravane in southern Laos, a pleasant two hour's drive from Pakse. Situated on the foothills of the Boloven Plateau, the resort consists of simple chalets overlooking the Se Set River and waterfalls in open jungle. Nearby are tribal villages which can be visited by elephant, which you rent at the resort.

Closer to Vientiane is the economically-priced **Lao Pako Resort** ((21) 216 600, advertised in handbills posted all over Vientiane. Located about 55 km (34 miles) from the capital, the small resort consists of simple thatched chalets on the banks of the Nam Ngum River, surrounded by forest. Getting there involves a one-and-a-half-hour drive to Somsamai, followed by a 25-minute motorized longboat trip up the Nam Ngum.

Sporting Spree

For action sports like paragliding, water skiing, or even windsurfing, it is better to head to Thailand or Malaysia; but that said, some adventure sports can be enjoyed in Vietnam. **Canoeing** amongst the massive limestone karst formations and 3,000 islands which rise almost perpendicular from the green waters of Ha Long Bay is an increasingly popular — and certainly the most adventurous, if not the best — way to see this remarkable work of nature. Fully accompanied trips including meals and all other arrangements can be organized through **Sea Canoe International** ((66-76) 212 252 FAX (66-76) 212 172 E-MAIL actioninfo@seacanoe.com WEB SITE WWW.seacanoe.com, PO Box 276, Phuket 83000, Thailand, a company now looking at river canoeing in Laos as an exciting new venture.

In the quickly developing resort of Nha Trang in the south central region, the offshore islands offer a variety of **scuba diving** opportunities. Several dive companies have sprung up in town, taking divers to the surrounding islands.

Vietravel ((58) 811 375 FAX (58) 811 374, 88 Tran Phu Street, Nha Trang, runs dive trips to a nearby island about 60 km (37 miles) away, where they have a guest house. **Ana Mandara Resort** ((58) 829 928 has its own dive shop at 86 Tran Phu Street. Also recommended is the **Blue Diving Club** ((58) 825 390 FAX (58) 824 214 at the Coconut Grove Resort, 40 Tran Phu Street.

In Da Nang, the **Furama Resort** ((511) 847 888, 68 Ho Xuan Huong, Bac My An, offers scuba diving, with a dive shop operating from the beachfront hotel at China Beach.

Visitors to the highland resort town of Da Lat, just an hour or two from Nha Trang, will see one of Vietnam's most beautiful golf courses and the first championship course in the country, the **Da Lat Palace Golf Club** ((8) 823 0227 FAX (8) 822 2347. The cool weather and early morning mist covering the greens makes playing on this course an almost mystical experience and for any golfer; a visit is highly recommended. Guests of the Sofitel Da Lat Palace Hotels or the Da Lat Novotel have automatic access to the club.

Another golf club to try in Vietnam is the Ocean Dunes Golf Club ((8) 824 3749

in Phan Thiet, adjacent to a Novotel Phan Thiet by a beautiful beach.

One of Vietnam's great adventures and the way to see the country first hand is to take to the road by bicycle. Several companies offer escorted Vietnam **cycling** tours. Intrepid Travel ((61 3) 9416 2655 FAX (61 3) 9419 4426, in Fitzroy, Victoria, Australia, 3065, offers a 14-day, low-impact, back road tour from Hanoi to Saigon, with a maximum of 10 people per group. Participants bring their own bicycle, have days off for relaxation, are followed by a backup vehicle, and cover the Hue-Hanoi sector by train. British Exodus Adventure ((44) (181) 675 5550 FAX (44) (181) 673 0779 E-MAIL sales@exodustravels.co.uk, whose head office is at 9 Weir Street, London SW12 OLT, offers a similar 14-day Saigon-Hanoi trip on paved roads, also with a backup vehicle and rest days, as well as a private boat trip to Cat Ba and Ha Long Bay (for groups of six to eighteen participants). The company has offices worldwide.

OPPOSITE: A minority woman tends her farm in Vietnam's central highlands. ABOVE: Elephant-riding through the jungle, from the That Lo Resort in southern Laos.

The Open Road

Some outstanding road travel awaits visitors to Indochina, especially those looking for off-the-beaten-track diversions. However there are some things to keep in mind. First, road conditions are quite different from in Europe or the United States. Distances have little to do with the time it takes to travel. Stick to driving 40 to 50 kph (25 to 30 mph) as a fair rule of thumb if you want to arrive in one piece and unfrazzled. Almost all hire cars come with a driver, which is by far preferable to driving a car by yourself. The only real freedom comes when you rent a motorcycle.

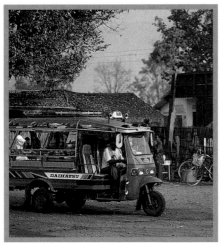

With Cambodia's current state of lawlessness, I would recommend (as do the foreign embassies) that all travel in this country be done by plane, except in the immediate vicinity of Angkor Wat. It is still dangerous to roam about the countryside, especially after dark.

Laos has some long distances to cover, especially in the south, where roads are gradually being upgraded to drivable standards. By 1999 they should be paved all the way to Pakse. A good day's drive is the route from Vientiane to Luang Prabang — an interesting run that cuts through the jungle- covered mountain ranges of central Laos before reaching the ancient royal city.

While it is possible to tour all of Vietnam by car, and some do, it involves a lot of driving and I would recommend breaking the journey up with flights and train travel, then driving to one or two particular areas for leisurely exploration.

Indochina's big drive is the epic journey along Vietnam's National Highway 1 from Hanoi to Saigon or the other way around. Vietnam's main artery is in good condition and, for the most part, remains uncrowded. While heavy truck traffic is increasing, there are still plenty of patches where the heaviest traffic will be a lone oxcart. For much of the way, the road passes directly through farmlands and small villages, where roadsides are lined with sheets covered with *padi* (rice) and other produce drying in the sun.

From the somewhat arid plains of southern Vietnam the road heads north past spectacular beaches before hitting the mountains at the Huy Van Pass, followed by the spectacular descent into the greener vistas of northern Vietnam, where it is not just the climate that cools, but also the atmosphere and attitude of the people.

With planning, it is possible to enjoy overnight stays in good hotels, often near spectacular beaches, such as the Victoria Hotel, near the old Cham outpost of Phan Thiet. You can follow this up by enjoying a more rigorous nightlife in Nha Trang staying in one of the city's numerous

hotels, or the quiet and beautifully otherworldly beach at Doc Lic, about an hour's drive (40 km or 25 miles) north of Nha Trang. Then head inland to the Imperial city of Hue and take the long road north to Hanoi. The Hue-Hanoi sector is probably better covered by train, the 12- to 16-hour drive not having many sights or facilities to enjoy. National Highway 1 can be covered by rental car, motorbike, bicycle or the less expensive Open Bus (see BACKPACKING, below).

One of Vietnam's best routes is the northwest mountain road that draws a giant loop from Hanoi west to Hoa Bin, then to Son La and out to the historic Dien Bien Phu. From Dien Bien Phu, the road proceeds across to Lai Chau then down to Sa Pa and the border town of Lao Cai and up to Bac Ha, before returning to Lao Cai and back to Hanoi. Although you can easily drive this loop as a five- or six-day tour, it is better to set aside a couple of extra days for exploring. Time the trip to be in Sa Pa on Saturday for market day and then in Bac Ha for its spectacular Sunday market, when all the tribal folk from surrounding mountains come to town dressed in their Sunday best. The trip is best made by

jeep or motorcycle, as some of the mountain roads are still a little rough. Hotels on this route are almost all of the rustic variety, with the exception of the new Victoria Hotel in Sa Pa.

In the South, the Mekong Delta can be explored by car, and in fact many tours offer just that. For far more comfortable exploration, for US$25 take the air-conditioned hydrofoil from the Saigon River jetty opposite the Majestic Hotel, at the end of Dong Khoi Street in Saigon, for a four-hour trip to Can Tho. From there take day tours by boat to surrounding areas of interest.

Backpacking

When thousands of youths take to the road across Asia, they create paths that quickly develop into well-followed trails, and eventually the upmarket tours of the future. Small backpacker beach

OPPOSITE: Siem Reap boys TOP playing in the river for their afternoon bath. BELOW: Southern Laos bus station is the town's sole transport base. ABOVE: Hue river ferry used by local women en route to market. OVERLEAF: Vung Tau beach is wide open during week days.

destinations often transform into resort towns: Bali's Kuta Beach and Thailand's Phuket provide good examples. In an adventure region like Indochina, backpacking is the norm rather than the exception; although in Laos, where mainstream tourism is still so new and destinations are opening all the time, the well-heeled and the penurious are often found sharing the same transportation, new destinations, and sometimes even the same accommodation.

In Vietnam, the backpacking trail is very well established, and the Vietnamese, not slow to recognize business opportunities, have provided plentiful low-cost services, including transportation. One of the best is the Open Bus route from Saigon to Hanoi with its comfortable, new, air-conditioned buses operating along National Highway 1. Daily routes ply between Hanoi and Hue, Hue and Hoi An (a big backpacker destination), Hoi An and Nha Trang, Nha Trang and Da Lat, and Da Lat and Saigon. For an eminently reasonable US$39, the ticket covers each sector of the route, or alternatively, it is possible to buy a ticket for just one (or more) sectors. Passengers can also disembark anywhere along the way, and then take local transport to the next embarkation point where they can rejoin the bus. Bookings are required only a day or less in advance which makes it very convenient. It can also be interspersed with train travel, notably for the overnight sector from Hanoi to Hue, where a soft sleeper is far more desirable (albeit more expensive) than an overnight bus ride. The Open Buses make several stops a day, at sites and good local lunch spots, making each leg of the journey into a tour of excellent value. Tickets for the bus can be bought at backpacker cafés and travel agencies in Hanoi or Saigon, or in downtown Hue, Da Lat, Nha Trang and Hoi An.

One word of warning. It is best to get off quickly when arriving at a destination. The budget market is an extremely competitive business and passengers are subjected to dreadful talks on the desirability of certain hotels and such, as purveyors try to drum up more business. Make sure you have a booking or the name of a good hotel and rent a cyclo or a taxi. In Hoi An, I spent an irritating hour being driven around to a dozen hotels while the driver refused to stop at any which weren't on his pay list.

With some guesthouses costing not more than a dollar or two a day for a shared room, low-end travel can work out to be remarkably inexpensive, with overall costs somewhere around US$5 or so a day. Traveling with a friend always works out more cheaply, as without the travel agent's profits included in the rate, most double occupancy rooms cost the same or only a small fraction more than for single occupancy.

A meal of market food costs less than a dollar, and backpacker cafés offer traveler's fare of pancakes as well as diluted versions of Chinese, Vietnamese, Laos, Khmer and Indian food for a few dollars a meal. Those willing to hunt will find better quality food in less popular restaurants not frequented by backpackers.

Laos is a different story. Being a much newer destination, the transportation infrastructure is less developed, and unless one is traveling on a group tour, or by rented car, public transportation is the only option. This means local buses or heavy trucks converted to buses with seats in the back — often a colorful, fun

and quite uncomfortable experience. Nevertheless, a backpacker trail has already developed and even in laid-back Laos, local transportation suppliers are quick to work out what tourists want (and what they are willing to pay), making it possible to get to major destinations without too much fuss or even without speaking any of the local language.

Cambodia too is at present very much a backpacking destination, once you get away from the environs of Angkor Wat. Many backpackers take the much-maligned speedboat service to Siem Reap. Why its bad reputation? Apart from being dangerously overloaded, a ride of similar distance and price (US$25) in Vietnam would buy a comfortable padded seat in an air-conditioned hydrofoil, while the Cambodia ferry has hard wooden seats, not to mention a thrill of danger I experienced as the driver zoomed past a suspicious-looking boat checkpoint in the middle of Tonle Sap Lake. With great relief, I flew back to Phnom Penh.

Sihanoukville to the south of Cambodia is a newly-developing destination of beaches and small islands. At present visitors are mainly confined to a few intrepid travelers and numerous

OPPOSITE: Central Vietnam. Pedicabs and bicycles are a popular form of transport in Hoi An. ABOVE: Siem Reap colonial buildings still stand in the center of the small town.

international workers, who live there and are used to dealing with the strained political situation. It is wise to check the current status in Phnom Penh before traveling.

Living It Up

Ritzy hotels, nightclubs, discos, and restaurants are not really a highlight in Indochina, but that said, a few deluxe days wallowing in luxury are eminently possible as are a few nights of reveling or gambling in reasonably ritzy casinos.

EXCEPTIONAL HOTELS
Many of Indochina's fine hotels are leftovers from an earlier, more elegant age, especially in Cambodia and Vietnam's northern and southern capitals.

Several of these onetime bastions of class have been renovated and lovingly restored to their former grandeur. For a price, the ultimate in elegance can be yours. Top choices include the **Hotel Sofitel Da Lat Palace** ((63) 825 444, at 12 Tran Phu Street, Da Lat, Vietnam, a truly divine establishment with spacious, elegantly furnished rooms to kill for. If you stay in only one grand hotel, it should be this one. The empty corridors seem to echo with ghosts from the days when dressing for dinner was *de rigueur* and ballrooms came alive every Saturday with a full orchestra.

Cambodia has several marvelous hotels, and no less grand is the old **Hotel Le Royal** ((23) 981 888 FAX (23) 981 168, at 92 Rukhak Vithei Street in Phnom Penh, which in its heyday played host to "globetrotters and adventurers, writers and journalists, royalty and dignitaries" from around the world. Later it served as home for journalists in the early 1970s before it became involved in the Khmer Rouge takeover. Part of the film *The Killing Fields* was made here. Equally grand is the lovely **Grand Hotel d'Angkor** ((63) 963 888 FAX (63) 963 168, 1 Vithei Charles de Gaulle in Siem Reap, first opened in 1929 for the "wave of

travelers for whom the Angkor Temples were an obligatory stopover." Less grand but no less evocative is the colonial **Renaske Hotel** ((23) 722 457 FAX (23) 726 100, 40 Samdech Sothearos Street in Phnom Penh, opposite the Royal Palace, an establishment with a long history. The marvelous garden filled with fragrant frangipani and palms is best enjoyed with a long and icy drink on the terrace in the evening, or over breakfast in the cool of the morning.

Back in Vietnam, in the resort town of Nha Trang, the **Bao Dai Villas** ((58) 881 049 FAX (58) 881 147 stand on a hilltop overlooking the sea. The villas each have several large and comfortable rooms, and a pleasant, if not completely elegant,

restaurant overlooks the sea. Also in Nha Trang, the very resorty **Ana Mandara Hotel** ((58) 829 829 FAX (58) 829 629 on Tran Phu Boulevard offers international standards of luxury and comfort near its own well-groomed beach.

Other attractive hotels include the old Saigon hotels that became so well known during the Vietnam War. Run by government tourist organizations, mostly Saigon Tourist, the standards are more Vietnamese than international, but this is what imparts a deliciously Vietnamese experience. My personal favorite is the charming and sprawling **Rex Hotel** ((8) 829 6043 FAX (8) 829 6536, at 141 Nguyen Hue Street, where the staff wear Vietnamese garb, the rooms are kitschly comfortable, and the rooftop garden is the best in Saigon. Down by the Saigon Riverfront, the tastefully redone **Majestic Hotel** ((8) 829 5512 FAX (8) 829 5510, at 1 Dong Khoi Street, offers grand deco-style rooms (don't forget to ask for the Western furnishings — they are far more attractive and in keeping with the hotel style) and a pleasant rooftop garden overlooking the river. The somewhat cold **Continental** ((8) 829 9201 FAX (8) 824 1772, at 132–134 Dong Khoi Street, lacks good management, but the rooms and garden restaurant have retained an amount of charm.

Phnom Penh 's Le Royal Hotel is arguably the best address in town.

In Hanoi, the **Sofitel Metropole** ((4) 826 6919 FAX (4) 826 6920, at 15 Ngo Quyen Street, has been the top address for almost ten years with excellent bars and one of Hanoi's best continental restaurants. Its new rival, the **Hanoi Hilton** (ACCOMMODATIONS, page 270 in TRAVELERS' TIPS), near the Hanoi Opera, should offer the best in ambience and service when it opens, some time late in 1998.

In Vietnam, the new **Victoria Hotels** are a French three-star chain bringing

much-needed international standards to secondary destinations, the first hoteliers brave enough to venture it. Their first, very pleasing, hotel opened in the French mountain retreat of Sa Pa ((20) 871 522 FAX (20) 871 539. A second is located at the beach resort of Phan Thiet ((62) 848 437/8 FAX (62) 848 440 on Vietnam's southern coast, about 200 km (125 miles) north of Saigon. Two more are located on the Delta — the first in the interesting market town of Can Tho and another in the Buddhist pilgrimage center of Chau Doc. Contact their Hanoi office ((4) 933 0318 FAX (4) 933 0319, 33 Phan Ngu Lao Street, Hanoi, for more information.

Laos's Luang Prabang, while it has no international chain hotels, has the charming and popular **Villa Santi** ((71) 212 267 on Sakkarine Street; the **Phousi Hotel** ((71) 212 192 or 212 717 FAX (71) 212 719, on Setthathirat Road; and the tiny **Le CaLao Hotel** ((71) 212 100, on Manthatourath Road, whose four top rooms overlook the Mekong.

The imposing **Sofitel Cambodiana** ((23) 426 288 FAX (23) 426 392, at 313 Sisowath Boulevard, dominates Phnom Penh, it is almost a city in itself, while moored behind it is the floating

Naga Theme Park — a disco, casino and entertainment center. The **Intercontinental Hotel Phnom Penh (** (23) 720 888 FAX (23) 720 885 E-MAIL phnompenh@interconti.com, recently opened a shining new presence there.

EXCEPTIONAL RESTAURANTS

VIETNAM

Pleasant bars and restaurants are found throughout Indochina, with the majority of good bars and eateries concentrated in bustling Saigon and Hanoi. Some of Hanoi's restaurants worth a visit include the **Mediterraneo** (Italian food) **(** (4) 826 6288, at 23 Nha Tho Street, and the pleasant and only slightly tacky but sweet **Le Cyclo Bar and Restaurant** (Vietnamese food and barbecue) **(** (4) 828 6844, at 38 Duong Tran Street. **Il Padrino (** 828 8849, at 42 Le Thai To Street, serves quality Italian food, and the marvelously quirky Spanish-style **Miro (** (4) 826 9080, at 3 Nguyen Khac Can Street, has a decor exceeded only by the quality and inventiveness of the cuisine. For an espresso coffee and delicatessen sandwich in a pleasant villa garden, try **Au Lac (** (4) 257 807, at 57 Ly Thai To Street opposite the Metropole Hotel.

In Saigon, the intimate **Augustin Restaurant (** (8) 829 2941, at 10 Nguyen Thiep District 1, enjoys the reputation of having the best French food in town, and **La Goulue (** (8) 836 9816, at 197 De Tham Street, in the center of backpacker territory, offers French cuisine at a budget price. A quirky treat is to eat in **Madame Dai's Bibliotèque (** (8) 823 1438, at 84A Nguyen Du, District 1, a Saigon institution that has been going almost as long as the *Doi Moi* open door policy has been bringing tourists to Vietnam. Her set menu French cuisine served in her library has had people talking for years.

CAMBODIA

Phnom Penh restaurants are alive and well, and in addition to the very ritzy restaurants in the upmarket hotels, there are dozens of places to sample around town. The **Foreign Correspondent's Club**

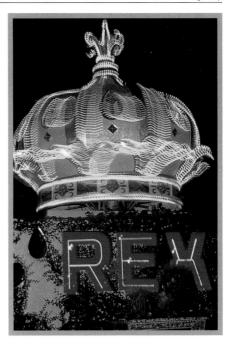

((23) 427 757 FAX (23) 427 758, 330 Sisowath Quay, offers big comfortable chairs, a view of the river, great beer, and wonderful food, possibly the best in town.

Le Deauville ((23) 801 955, at Wat Phnom, serves charcoal-grilled fish in pleasant surroundings near Phnom Penh's namesake wat. The Thai **Chao Praya Restaurant** caters to tour groups and individual diners in a charming renovated French villa on Norodom Boulevard. **Le Casablanca (** 805 816, on Rue De France north of Wat Phnom, offers charcoal grills and North African cuisine.

LAOS

Decent restaurants in Vientiane include **L'Opera (** (21) 215 099, near Nam Phu Fountain, an Italian Restaurant reputed to be the best in town The very pleasant **Le Vendôme (** (21) 216 402, in Wat Inpaeng Street behind Wat In Paeng, serves French and Lao food.

OPPOSITE : Saigon's quirky Maxim's Restaurant TOP offers an eclectic menu and a taste of old Saigon. Da Lat BOTTOM is the south's source of roses and fruits. ABOVE: Lit by a giant revolving crown, the roof garden at Saigon's Rex Hotel is an enduringly popular dining and drinking spot.

In Luang Prabang, I recommend **Auberge Duang Champa** ((71) 212 420 FAX (71) 212 420, adjacent to the Nam Khan River on Kingkisalath Road, where set-menu meals come amazingly cheaply. They also offer à la carte dishes and a reasonable French wine list. The **Villa Santi** ((71) 212 267 on Sakkarine Street offers authentic French and Lao cuisine prepared by the daughter of the last king's chef, while the highly recommended **Nam Kharn Garden Restaurant** ((71) 212 476, on Kingkitsalath Road overlooking the Nam Khan River, offers delightful views, a friendly and pleasant atmosphere, and very reasonably priced Lao meals with some Western additions.

NIGHTLIFE

VIETNAM

Not-to-be-missed bars in Hanoi include the northern version of **Apocalypse Now** ((4) 971 2783, at 5C Hoa Ma, which gets busy late in the week, and the **Stone Elephant** ((4) 828 4545, at 2 Cua Dong with live jazz on Friday nights. The **Art Café** ((4) 828 0905, at 35 Luong Ngoc Quyen, offers "drinks and midnight snacks."

Saigon too is brimming with places to live it up after dark. Don't miss the **Globo Café** ((8) 822 8855, at 6 Nguyen Thiep, District 1, a hip little French restaurant, with "great food and good prices," that warms up after 10 PM Thursdays through Saturdays with a live jazz band squeezed into a corner. The **Q Bar** ((8) 823 5424, at the rear of the City Concert Hall, has been a perennial favorite in Saigon with expatriates and visitors alike, as has **Apocalypse Now** ((8) 824 1463, at 2C Thi Sach, District 1, where everyone heads for at least a drink and a look at this recreation of 1960s Saigon. The **Press Club** ((8) 929 1984, at 39 Le Duan, District 1, offers inexpensive drinks. All are welcome. The **Gecko Bar** ((8) 824 2754, at 74/1A Hai Ba Trung, is popular with expatriates. Try their specialty: gecko wine. And **Café Latin** ((8) 822 6363, at 25 Dong Du District 1, offers *tapas* and happy hours, and attracts a popular crowd.

CAMBODIA

Cambodia has reverted to dark nights and lawlessness since the days of UNTAC (United Nations Temporary Administrative Command), when there was a bar and restaurant on every corner

and a feeling of security. At present, people tend to keep indoors at night, unwilling to venture out. Those that do will find plenty of places to visit and a local crowd of expats who seem to thrive on the edginess. The local tourist rag *Bayon Pearnik* runs fun features with topics like "Keeping Your Head While Losing Your Wallet," providing amusing advice such as "hand over your valuables to the man with the gun" or "do put your hands in the air — it adds to the feel of being in a holdup situation." And "don't start making polite conversation, for example 'Ooh what a big gun you have… Is that a K59 or a Smith and Wesson?'" In other words, it's probably just like New York!

Those who do venture out at night will find giant discos like **Martini Pub**, at 402 Mao Tse Tung Boulevard, which advertise: "Lonely, bored, hungry? We have everything you need!" This dance hall is filled with young girls from the country and eager-looking men. **Heart of Darkness**, at 26 Street 51, is reminiscent of Apocalypse Now — a dark, boozy sort of place that runs hot all night with loud music and a pool table. It is popular with braver backpackers and NGO workers.

Adjacent to the Cambodiana Hotel is the **Naga Resort** floating casino — a converted ship that entertains the moneyed classes. Besides the casino, which is open from 2 PM until 6 AM, the Pirates Fun Pub and Disco is open until 3 AM, as is the karaoke lounge, and the terrace café stays open until 6 AM.

LAOS

Laos, ah Laos! Vientiane offers a few semi-clandestine pubs popular with gays, expatriates and the odd Lao girl, although the government frowns on too much decadent nightlife. What is open one week may well be closed the next, so keep your ears to the ground for the latest news. The **SamLo Bar** is quite popular, but once again sometimes it's open, sometimes it's closed. The **Fountain Bar** in Nam Phu Circle seems to be a popular if somewhat seedy spot with all sorts patronizing it. Popular sunset bars can be found strung out along the banks of the Mekong. Try the

OPPOSITE: Colonial architecture of the Grand Hotel faces Nha Trang's Beach Boulevard and the broad beach. ABOVE: Hanoi bar girls in one of the city's many drinking establishments.

ones farthest from town, past the Riverview Hotel.

Luang Prabang's most raucous nightlife may be enjoyed at the disco beneath the **Rama Hotel** where everyone turns up — young Lao couples, tourists and backpackers — for a look at unsophisticated but entertaining Laos nightlife.

Family Fun

In all my travels in Indochina, I have seen only one or two families traveling with children, although that is not to say it is impossible. With most travel still of the adventure variety rather than deluxe, it seems that most folk leave the kids at home. Nevertheless, Asians love children and delight in playing with and looking after them, whether they are their own or somebody else's. To find a suitable person to look after or entertain the kids for a few hours is no problem at all.

Any family brave enough to bring the kids will find that traveling together, even to remote parts, can bring another dimension to travel and help to make many new friends. Do bear in mind that children can get ill, picking up stomach bugs and the like, but a modicum of care should help to prevent any catastrophe. Doctors and excellent (foreign-run) medical services are readily available in the main towns.

While temple tours and endless guide-related explanations could prove tiresome to children whose threshold of boredom is somewhat lower than most adults, a relaxed walk in a wat, meeting young monks, could prove to be a delightful experience.

No doubt due to the fact that most Asians travel *en famille*, complete with aunts or grandparents, extra beds are readily available at a reasonable price all over Asia. Don't forget to take a ready supply of books and drawing equipment, basic medicines such as for diarrhea, mosquito repellant, and other necessities. Bottled water is readily available all over Indochina, as are canned soft drinks.

Cultural Kicks

In a region where most tourist attractions are cultural rather than hedonistic, just exploring Indochina can be a cultural kick in itself — a kind of absorption of the delights on offer. Dance and drama, museums and music, wats, temples, and often palaces abound in all the major cities, especially in Laos and Cambodia, each crammed with exhibits of stone statuary, royal regalia and Buddhas that would be worth a king's ransom on the open market (which is why so many statues have been removed from ancient sites).

DANCE

Music and dance are part of the cultural spectrum one could experience during an Indochina visit. Cambodian and Lao dances are more subtle and stylized than those of the Vietnamese. The elaborately-costumed court dances share the same Indian influence notable in both Javanese and Thai dance. In Cambodia especially, dancers are stitched into their form-fitting, heavily-brocaded garments, and wear jeweled headgear. Movements of the face and hands tell the story and

OPPOSITE: Mekong Delta hammocks are enjoyed by boys of all ages. ABOVE: New Saigon traditional opera is a colorful mix of Chinese and Vietnamese influences.

drama of the choreography, generally representing variations on the classic Indian *Ramayana* theme. In fact, every January a *Ramayana* festival is held on the November full moon at Angkor Wat — a splendid three-day festival where invitees from Laos, Malaysia, Java, and Thailand try to outperform each other with magnificent versions of this ancient Indian tale of the triumph of good over evil.

Folk dancing throughout Indochina shares similarities and manifests subtle differences. With so many ethnic minorities hidden away in the highlands, it all depends on the tribal groups performing. Dances of the northern tribal groups resemble those of Vietnam, Laos, Southern China and Thailand, as the communities are spread across that whole region. Coming from completely different origins, the dances of southern tribal groups offer something quite different.

Unfortunately, unless you go to an ersatz hotel cultural show, or catch the odd performance during a festival, the chances of seeing authentic dance are pretty slim. However, several hotels stage regular cultural shows with buffets, and the quality is good. The Villa Santi in Luang Prabang, for example, features regular shows of traditional Lao dance and music, as does the Hotel Lan Xang in Vientiane.

In Cambodia, the impressively renovated Grand Hotel d'Angkor in Siem Reap features nightly performances by local dance troupes, and the highly-regarded National Cambodian Dance Company occasionally stages performances in the forecourts of Angkor Wat.

MUSIC

One fascinating aspect of Vietnamese traditional music is the unique and quite bizarre range of musical instruments. In the north, many stylistic aspects of traditional music and dance can be traced back to an early Chinese influence, with a similar five-note pentatonic scale and orchestras of up to 40 musicians

playing the sort of instruments you'd see in China.

But the moody, melodic base of Vietnamese music is provided by an instrument which is very much their own — the *danbau*, a single-string zither with a willowy tuning rod at one end which, caressed by the musician's fingers, controls the tone and pitch of each note. In addition to reed flutes, gongs, and mandolin-type stringed instruments, they also play a bamboo xylophone which hangs like a curtain before the musician. Then there's another bamboo instrument for bass accompaniment in which the hands are clapped at the mouths of thick, graduated tubes very much like the pipes of an organ; the action produces pulses of air which provide a series of low, resonant burps.

These instruments provide the basis of folk, classical, chamber, and theatrical music and dance which are staged throughout Vietnam, in the municipal opera houses of Hanoi and Saigon, at conservatories and cultural centers, and at special hotel cultural shows.

Vietnam's cosmopolitan heritage means that Western music also resounds from municipal halls, dance clubs, and bars around town. Ballroom dancing is very popular in Vietnam — a bizarre anachronism maybe, but pursued by hundreds of couples in special dance halls complete with VERY LOUD, amplified music and five-piece orchestras. At one end of the musical spectrum, you can eat to the music of Chopin at the Piano Bar in Hanoi, or enjoy a student chamber recital at a conservatory right alongside the Citadel in Hue; and at the other, drop into clubs in Hanoi or Saigon and marvel at their rock and pop music.

In Laos and Cambodia, mainstream traditional music shares a common heritage with Thai music, using semicircular gamelan and gongs and lilting hill-tribe bamboo flutes.

Tribal boy wears a traditional headdress in northern Laos.

While classical music exists in Laos, it is the folk music that holds the heart of the Lao people, as exuberant as it ever was, and serving as the country's popular music. One day I was on a local long-distance bus with Lao music played over the loudspeaker, so pleasant and fitting for the situation. Unfortunately, a group of tourists decided to change the status quo and imposed some jarring Western music on the bus population. A spell was broken. Being Laos, nobody really reacted verbally, but there was a subtle shift in the harmonious atmosphere which had existed just minutes before.

PUPPETS

Another compelling craft is Vietnam's inimitable wooden water puppets, which

have moved from their village origins (where they are operated on the village pond, usually near a small temple or shrine) to the big city, where most visitors will be able to catch an indelible performance. In Hanoi, a performance can be seen in the **Kim Dong Theater** ((4) 824 9494 or (04) 825 5450, 57 Ding Tien Hoang, or at Water Puppetry House, 132 Truong Chinh Street, where a troupe founded by Ho Chi Minh performs twice weekly.

In Cambodia the ancient art of shadow puppetry is still alive if not well — another ancient art that suffered at the hands of the Khmer Rouge. Shadow puppet performances, as well as folk music, are often held during country weddings and festivals.

MUSEUMS

Among the best museums to visit is, in Laos, the **Royal Palace Museum** in Luang Prabang, once the palace of the royal family. This ornately-decorated palace is filled with precious Buddha images and possessions of the last king. In Cambodia, the **National Museum** in central Phnom Penh, adjacent to the Royal Palace, is beautifully designed with Cambodian Khmer-style architecture and is filled with precious stone statues from Angkor Wat and other significant Khmer sites across the country.

In Vietnam, the **Fine Arts Museum** at 66 Nguyen Thai Hoc in Hanoi is well worth the visit. On the outskirts of Hanoi at Dich Vong in Cau Giay there is the imposing, newly-completed and absorbing **Ethnographic Museum** (836 0350 with displays that encompass a large proportion of the country's 54 ethnic minorities. The exhibits are well put together and give a reasonable picture of the lives and cultures of the country's various minorities. In Da Nang, the **Cham Museum** (open from 7 AM to 6 PM) exhibits Cham statuary from the surrounding sites, many of which were bombed to extinction during the war. It is located at the southern end of Bach Nang Street.

Shop Till You Drop

Glitzy shopping malls with the latest designer fashions are not what Indochina is about — save all that for a side trip to Singapore, Hong Kong, or Bangkok. As in China, the socialist revolution stifled or deliberately crushed the cultural diversity of Vietnam and Cambodia, and to a much lesser extent, Laos, replacing traditional arts and crafts with politically correct propaganda.

However, a reemerging national pride and a lessening of Draconian policies — due in part to *Doi Moi* — is giving birth to a lively resurgence of crafts production all over the region, as traditions are being resurrected. Beautifully produced handicrafts, gems, silverware, ceramics, textiles, lacquerware, antiques and, in Vietnam, exciting new art works from emerging artists (see SPECIAL INTERESTS, page 65), are available in abundance. Many pleasant hours can be spent not only shopping, but learning about Indochinese cultural mores.

In many cases, it is necessary to list individual shops here, because although crafts are available in abundance they are not easy for foreigners to find, unless you do plenty of research and take a guide.

Vietnam has a wealth of good buys. The pale green Celadon ceramics so beloved of collectors are still being made in Hanoi using traditional methods. Where treasured old pieces can fetch hundreds of dollars on the international market, new pieces are very reasonably priced, selling for US$10 to US$20. **Celadon Green**, at 29 Dong Du Street in Saigon, offers a wide range of pieces from bowls, cups, and teapots to plates, jars, and urns.

The dusty old lacquered idols and statues found in temples, where centuries of incense smoke have added a patina of preciousness, are now being manufactured again, and although they lack the telling wear of age, some are beautiful. For a good selection try **Antika** ((4) 828 3583, at the Camelia Hotel, 13 Luong Ngoc Quyen in Hanoi; **Quang Minh** ((4) 825 1497 at 40 Hang Be, Hanoi; or **Heritage** on Dong Khoi Street in Saigon.

OPPOSITE: Monks relax TOP alongside a stone lion outside Wat Xieng Thong in Luang Prabang. BOTTOM: Carol Cassidy 's Vientiane studio creates handlooms that sell to the elite. ABOVE: A Saigon painter in Dong Khoi street studio.

Handmade glassware can be found at **East Meets West**, 24 Le Loi Street, District 1, Saigon. Glassware and exuberant hand-painted ceramic tea sets sell for around US$20 alongside numerous other Vietnamese-crafted goodies at **The Home Zone**, 41 Dinh Tien Hoang, District 1, Saigon. Magnificent gold and green glazed tiles are being made in Hue using the same methods as used centuries ago on the emperor's tombs.

Elegant platters of lacquer-covered bamboo, some finished in gold leaf, others in earthen shades of terra-cotta, are made in some small villages outside Hanoi. Sizes range from small to enormous, and the platters make an elegant design statement in a modern room. Several of the gift shops near Hoan Kiem Lake sell them, but check each piece before buying. There is no quality control, and some pieces will have flaws. **Heritage** at Dong Khoi Street in Saigon sells quality pieces. In Hue, fine traditional lacquerware can be found in a number of shops, but beware of the garish imitations found in Saigon. If you have a good eye, search out for yourself, otherwise try the very tasteful **Heritage Art Shop** ((54) 823 8438 at 53 Dong Khoi Street, whose owner has scoured the country for the best crafts.

Although many of the tribal cultures have been all but obliterated in Vietnam, due to a Vietnamization process, a new cultural awareness seems to have surfaced in the city, and tribal textiles and costumes are suddenly coming into vogue — no doubt also assisted by the rising tide of nongovernmental organizations. Shops in both Hanoi and Saigon are selling tribal textiles, whole outfits, and colorfully assembled models for tourist wear. Shops sponsored by nongovernmental organizations in Hanoi sell tourist-type, handloomed textiles, which are made especially for the tourist market and are pretty but worth little as collection pieces. Shops to try in Hanoi include **Quang Minh — The Culture of Vietnam Ethnic Groups**

((4) 825 1947, a little shop with many interesting things at 40 Hang Be Street in Hanoi's Old Quarter. A second shop with the same name is located at 44 Hang Gai Street ((4) 828 0509, also in the Old Quarter. The **Pan Flute** ((4) 826 0493, at 42 Hang Bac in the Old Quarter, sells minority clothes and accessories, while **Annam Crafts** ((4) 863 4551, at 3 Dinh Liet Street, sells traditional Vietnamese handicrafts.

For real quality textiles, it is necessary to head to Laos, where handwoven textiles are part of the lifestyle. All Lao women wear them as a matter of course — and by government decree, an effective way to keep the home industries turning. The diversity of weaves and patterns is nothing short of extraordinary, although, patterns are becoming increasingly standardized as the government tries to introduce a "national" mentality rather than a regional one.

Hand-spun silks and cottons are the main materials used, and the best are hand-dyed using traditional natural dyes from plants and trees, whose colors are so rich and yet subtle, they appear almost luminescent. Many pieces feature intriguing animistic patterns and motifs, portraying the beliefs of the hill-tribe artists who created them.

Even modern articles, which purists tend to scorn because they are chemically dyed, can be beautiful, and a visit to the weaving village of Ban Phanom just four kilometers (two-and-a-half miles) outside Luang Prabang will convince you that, while the dyes have changed, the workmanship and creativity haven't deteriorated.

As the vogue increases, so do the prices. It is not unusual to see a good example of an old handwoven piece sell for US$1,000 — an already established market price. Check the morning market in Vientiane for an absorbing delve into the world of textiles.

A Lao woman displays a handloomed woven skirt in Boloven plateau village, southern Laos.

Laos is also a good place to find examples of superb, finely-wrought, authentic tribal basketwork, as well as attractive new styles. Antique Chinese ceramics and porcelain can also be found in antique shops in both Luang Prabang and Vientiane.

In Laos the tourist market is quite well-catered to, with some very attractive antique shops in Luang Prabang 1169; and textile stalls too numerous to mention. Shops and galleries worth visiting include the wackily-named **Doris Jewelry** ((71) 218 800 extension 1169 — the name belies the quality of her items — run by Vivian Wang, in glittering (and air-conditioned) Lao Hotel Plaza, 63 Samsenthai Road. Her superb collection of textiles and antiques are worth spending on. The Lao Women's Union's **The Art of Silk** ((21) 214 308 has a marvelous collection of antique textiles in the upstairs museum and plenty of nice pieces for sale downstairs. **Kanchana** in That Dam District, 102 Samsenthai Road, has beautiful antique pieces and well-executed modern pieces which incorporate traditional designs in tone-on-tone colors. Within the spacious **Lan Xang Hotel** is a shop run by the charming Mr. Bounkkong Signavong (((71) 313 223), with more antique Lao textiles and other sublime pieces. **Satri Laos Silk** ((71) 543 874, at 79/4 Setthathirat Road Wat Mixay, is

also worth visiting. The studio of internationally-known American Carol Cassidy — a weaver who bills herself as having single-handedly "rejuvenated the Lao silk industry" — can be visited at her **Laos Textiles Studio** ((71) 212 123 FAX (71) 216 205 just off Setthathirat Road, where visitors can view her modern pieces using traditional Lao designs. Not to be missed is Vientiane's **Thanon Thalat Sao** — Morning Market — which stays open all day. It is crammed with textile stalls and some local handicrafts of varying quality. The number of stalls selling antique pieces (and imitation antique pieces) is expanding rapidly, due to escalating demand.

Cambodia is a little short on quality crafts, surprising for a country with such a long traditions in the arts. Presumably the country's top craftsmen did not survive the Khmer Rouge purges. In Phnom Penh, several art shops sell some fairly tired-looking souvenirs and silver work, but with the country's reputation, the current offerings prove disappointing. Coarse and roughly made, they come nowhere near the quality of older pieces which, when you can find them, are worth picking up.

Phnom Penh's large yellow neo-constructivist Central Market gives a fair indication of its contents. The heavy concrete walls hold a core section full of fine gold jewelry, and some silver and watches, but otherwise there's not much available except T-shirts.

The most exciting place to head for is the **Russian Market**, on Street 182 near Street 111. It is quite far from the downtown area, and it is best to take a motorbike taxi or even a taxi. Within the dim interior are antiques, textiles and an abundance of newer handicrafts in small stalls, interspersed with other stalls selling food, produce, fruit, tools and hardware, and piles of rice and spices — it all adds to the color. Also to be found in the Russian Market, and no doubt in other markets around town, are several stalls selling marvelous blue-and-white

china with exuberant floral designs. Hand-painted cups, saucers, plates and bowls are piled high, nesting in straw holders like some strange birds. Although we are not talking Wedgwood quality here, the rough execution is as eminently pleasing as is the price. A whole dinner set could be had for less than US$20, but transportation could cost a little more.

Cambodia is best known for its quality textiles, especially the fine old silk *ikats* that are becoming increasingly difficult to find. One ray of light on the Phnom Penh shopping scene is the expensive **Orient** ((23) 725 308 at 245 Quai Sisowath. This tempting shop has a tasteful collection of Chinoiseries.

In the dozens of stalls around Phnom Penh's **Central Market** are piles and piles of new handloomed silks and cottons which, while lacking the luster of the old ones, are still very attractive and quite collectible. Handwoven silk or cotton Khmer headwear — the distinctive red-and-white or blue-and-white checkered scarves worn by most rural Cambodians and the Khmer Rouge — sell for a few dollars and make fine gifts and souvenirs. Beware when buying, some are handwoven with polyester, which is not quite the same thing.

OPPOSITE: A boy in Central Vietnam sells trinkets on Lang Co beach. ABOVE: The Mekong Delta floating markets are one of the region's "must sees."

Festive Flings

Festivals in these very rural countries usually revolve around a religious event or an agricultural ritual. Festivals play an important part in the lives of the people of Indochina, and there are important ones happening somewhere in the region nearly every month of the year. Some dates, like the international New Year, are fixed, but most are lunar dates, shifting in accordance with the full moon, which makes it relatively easy to work out the date a festival will take place.

JANUARY
The **International New Year**, on January 1, is celebrated across Indochina as a national holiday. January 7 is Cambodia's **National Day**, which celebrates the fall of the Khmer Rouge. In mid-January, Laos observes the **Boun Khoun Khao**, a harvest festival celebrated in the villages, where ceremonies are performed to give thanks to the spirits of the land.

Falling sometime during December and January in Laos is the **Boun Pha Wet**, a festival based on temple celebrations with recitations of the *Jataka*, the story of Buddha's life. It is also a time when monks are ordained. The actual dates vary from village to village, to enable villagers to visit their friends and relatives during their celebrations.

FEBRUARY
On the full moon of the third lunar month (usually early February) **Magha Puja** is held, celebrated with particular fervor at Wat Phu in southern Laos. During this massive celebration, crowds descend on the ancient wat making offerings to the spirits and giving alms to parading monks. Bullfights and boat races are part of the celebrations.

Tet, the Lunar New Year, is celebrated by the Chinese and Vietnamese communities all over Indochina. Firecrackers are generally banned, making it a far less momentous occasion than previously. Vietnam shuts down

tight for five days and lesser celebrations occur in Laos and Cambodia. In mid-February in Vietnam, the **Lim Festival**, a folk singing festival, is held 30 km (19 miles) outside Hanoi in Lim Village in Bac Ninh Province, and traditionally-dressed folk from 49 villages converge in the tiny town.

MARCH AND APRIL
Cambodia celebrates **Women's Day** on March 8, with parades and floats in the main towns.

The **Boun Pi Mai**, Laos's New Year, is held on April 15, 16, and 17. It is best witnessed in Luang Prabang, where three days of temple offerings, elephant parades, and temple activity herald the

new year. A tip: wear old clothes and watch out for water fights; visitors receive no mercy when it comes to water buckets. Boun Pi Mai is a national holiday in Laos and lasts for at least three days. Cambodia's **Bonn Chaul Chhnam** celebrates the new year in mid-April with a national holiday, and its Independence Day is held on April 17, with parades in Phnom Penh.

MAY
Cambodia, Laos and Vietnam all celebrate **Labor Day** as a national holiday on May 1, and on May 6 Cambodian rural communities celebrate **Bonn Dak Ben** and **Pchoum Ben**, the Royal Ploughing Ceremony, which is an ancient Brahman ritual celebrating the beginning of the rice-planting season. May 9 is Cambodia's **Genocide Day**, held in memory of victims of the Khmer Rouge.

In Laos on the May full moon is **Boun Visahhabousa (Vesak** or **Waicak)**, when the country celebrates the birth, death and enlightenment of Buddha. Chanting and candlelit processions are held in the wats. Vietnam also celebrates Vesak Day, as does Cambodia. In Laos, **Boun Bang Fai** (Rocket Festival) — a pre-Buddhist rain and fertility rite — coincides with Vesak Day and is celebrated along with Vesak on the days of the full moon. One

The devout make offerings to Buddhist monks at Wat Phu festival in southern Laos.

of the best and most enjoyable festivals, it includes dancing, music, and parades of giant wooden phalluses. In the villages, bamboo rockets are shot into the clouds to bring down the rain.

May 19 is **Ho Chi Minh's Birthday** in Vietnam, which is a national holiday.

JULY, AUGUST, SEPTEMBER
In Laos, the **Boun Khao Phansaa** is celebrated on the July full moon, marking the beginning of Buddhist Lent — a quiet time when monks remain in the monastery. It is also the time for many ordinations in the wats.

In Laos, the August **Boun Kao Padap Dinh** is the time for cremations and offerings to the dead, and is not necessarily a sad time at all.

September 2 brings Vietnam's **National Day**, commemorating the 1945 Independence Proclamation by Ho Chi Minh.

OCTOBER, NOVEMBER, DECEMBER
On the October full moon, Laos celebrates the happy **Boun Awk Phansaa**, marking the end of the three-month monks' retreat, with **Lai Hua Fai** (the Festival of Lights). Lit candles are floated in small banana leaf boats on rivers and lakes. It is held in conjunction with the **Boun Nam Water Festival** and **Boun Suang Heua** (boat races) where longboat races are held in towns with rivers or lakes, especially Vientiane and Luang Prabang. In Cambodia, **Bonn Kathen** is a 29-day festival where town and country inhabitants bring new robes to the wats for the monks.

Cambodia celebrates His Majesty the King's Birthday on October 30 and November 1 with festivities including parades and fireworks on the river near the palace.

The popular **Boun That Luang,** the November full moon festival at Wat That Luang in Vientiane, represents the biggest festival of the year. Monks gather at Vientiane's largest wat to receive alms on the first day. Parades between Wat That Luang and Wat Si Muang, fireworks, and a massive candlelit circumnavigation of Wat That Luang accompany the carnival.

The November 9 Cambodian **Independence Day** celebrates the country's independence from France. On November 25 and 26 longboats race and fireworks pop during Cambodia's **Water Festival**. On the days around the November full moon the **Ramayana Festival** is held at Angkor Wat with three days of classical dance.

On **Laos National Day**, December 2, speeches, parades, and family fun honor the power of the people during this joyous public holiday celebrated across the country.

Short Breaks

In Southeast Asia, short breaks are generally the norm for Asian tourists, and reasonable packages often appear in the local papers. If one is vacationing in Bangkok or Singapore it is just a simple flight to one of Indochina's top destinations, a short break rather than an exhausting extended tour.

Top of the getaway list is the ultimately charming royal city of **Luang Prabang**, which is directly accessible by Bangkok Airways from Bangkok, Thailand, or by a short flight from Vientiane. The Indochinese equivalent of Bali's Ubud, Luang Prabang is

OPPOSITE: A giant Naga head stands amongst the stupas in Phnom Penh's Royal Palace.
ABOVE: Monks outside the gilded doors of a chapel at Wat Xieng Thong in Luang Prabang.

becoming recognized amongst artists and the romantically inclined as a place to be. Recently acknowledged as a UNESCO World Heritage Site, the small town sits at the confluence of the Nam Khan and the Mekong Rivers. With its palm trees and golden spires, its uncrowded streets lined with colonial architecture, its ancient wats punctuated by the bright saffron robes of wandering monks, and its big, red, Mekong sunsets, Luang Prabang is a romantic setting designed to charm. In the shadow of Mount Phu Xi, with the tiny That Chom Si and its bright, golden stupa and views of the town, Luang Prabang is a delight. Add to this a few charming boutique hotels and some decent restaurants, not of the Michelin-star variety but pleasant enough, and you have the all ingredients for a inspired short stay — so good in fact that some visitors want to stay for months.

Vietnam's northern capital, with its leafy tree-lined streets and dun-colored colonial buildings, also has a sedate charm that invites wandering. Days can be spent exploring the many museums and markets at a leisurely pace, amply provided for by the bicycles and cyclos that still pedal the streets. **Hanoi** is particularly pleasant in winter, when cold temperatures make it necessary to bundle up in warm clothes and the atmosphere takes on a strangely disconcerting, "Europe in Asia" feel.

For a beach-resort vacation, **Nha Trang** ranks as a surfer's paradise, or possibly the Pattaya of Indochina, minus the bar girls. Dancing girls aside, Nha Trang offers plenty of entertainment, including excursions to nearby islands, diving and other water sports. The long, sandy, beach fronts the town, and numerous old colonial-style hotels are still standing even as the new highrise hotels are taking over. For a relaxed respite from just a about everything, Nha Trang is just what a lot of people fancy — I met more than a few people who were spending their whole vacation lazing on the beaches.

Angkor Wat is accessible by direct flight from Bangkok, and even day tours (not recommended) are available for those in a hurry. It is better to spend at least two or three days there, allowing time to explore in a leisurely manner.

Galloping Gourmets

The proletarian revolution which followed the war reduced the cuisines of Vietnam and particularly Cambodia to the level of basic sustenance. Cuisine was considered a bourgeois decadence. This philosophy, compounded with the austerity and social leveling the Communists brought in, spelled disaster for food lovers. However, the return of tourism and the demand for something more than boiled rice and vegetables has revived the region's restaurant industry, and Vietnamese cuisine especially is as good as it ever was.

VIETNAMESE CUISINE
A cuisine with a strong Chinese influence, Vietnamese cooking is somewhat less varied than that of its northern neighbor — covering probably 500 different dishes compared with some 2,000 in the Chinese culinary compendium. Yet it is distinctive and, with its strong reliance on natural ingredients, is perhaps more delicate and subtle.

The use of abundant amounts of fresh greens lift it to a higher plane, giving the food new meaning. Salad leaves include lettuce, mint, basil, and other unfamiliar herbs which you pluck and add to your dish.

Indochinese food centers on the ubiquitous noodle soups, *pho*, pronounced "for" in Vietnam and "fur" in Laos. The most popular are *pho bo* (beef noodles) and *pho ga* (chicken noodles), and great steaming bowls of these soups are both heartwarming and nourishing — the comfort food of Indochina.

Besides language and architecture, the French left behind a wonderful legacy of bread-baking, and Indochina is one of the few, if not the only, part in Asia where you can find decent bread.

Every morning street stalls in all the main towns, and even in country villages, are piled with fragrant fresh bread, ready to eat, accompanied by patés or fresh fried eggs and cups of fragrant coffee.

OPPOSITE: Smoke from cooking fires adds a touch of romance to a Luang Prabang evening scene. ABOVE: This French pasty shop in Hanoi has served the likes of actress Catherine Deneuve.

Ironically, possibly as a move to please their customers, some hotels are serving softer, processed breads rather than the crusty, French-style baguettes of the streets; fortunately, there is still a minority who maintain tradition.

Crisp and fresh, baguettes are especially good in Saigon with rough paté, or better yet, with loads of fresh Vietnamese butter and Da Lat strawberry preserves (possibly the world's best — it comes studded with whole strawberries floating in a sea of sticky strawberry toffee). Two of the most pleasant places to enjoy this delicious breakfast, accompanied with a cup of fragrant *café filtré* and fresh fried eggs (not the factory variety), are the Roof Garden of Saigon's Rex Hotel in the cool of the early morning, or the Givral Café facing the newly-renovated Opera.

It's another delight altogether to experience Indochinese coffee. Although freeze-dried Nescafé is flooding the region — considered to be quite the sophisticated thing to offer — you can still get the traditional beverage in most cafés and restaurants. It's locally-grown coffee filtered onto an inch or so of sweetened condensed milk, then stirred into a mixture which has the color and texture of *padi* (mud). It's unusual, but it provides a deliciously powerful kick-start to any day.

Vietnam's tasty street food can be enjoyed across the country. One of the most well-known and ubiquitous dishes is *cha gio*, or fried spring rolls. The paper-thin, rice-flour wrappers contain a mix of prawns, minced pork, bean sprouts, cellophane noodles, and vegetables, and appear in various forms throughout Vietnam. The rolls are wrapped in salad greens and dipped in fish sauce before being eaten with chopsticks (foreigners are allowed to eat with their fingers).

Cold, white rice noodles known as *bun* are served in a variety of ways. One of my personal favorites is *bun cha* — freshly-barbecued slices of pork or pork patties served with a delicious, slightly spicy dipping sauce, *bun* and salad. Other varieties include *bun bo*, which comes

served in a large bowl with *bun*, chopped spring roll, beef slices, bean sprouts, fried garlic, and a spicy sour sauce. It is quite possible to eat this delicious meal for days on end without ever tiring of it.

Hanoi has some of the best street food in Asia. In the heart of the Old Quarter, women sit at tiny tables on the pavement huddled over a charcoal fire, cooking their specialty — the one dish they know how to make well. The customers sit on doll-sized chairs while they sample her fare. In this eat-and-run society, the women can handle a large turnover in an evening, and even the well-to-do will stop by for a plateful of their favorite snack. Some of these roadside restaurants will serve up piles of freshly cooked red crabs, but beware of the tourist markup.

A unique Hanoi specialty that dates back to the city's early days is *cha ca*, a dish available at only one traditional restaurant in the Old Quarter, called **Cha Ca La Vong** ((4) 825 3929 located at Cha Ca Street. They now run a newer branch at 107 Nguyen Truong To Street ((4) 823 9875. The quirky old restaurant in Cha Ca

Street serves nothing else save this inimitable fish cooked at the table over a charcoal brazier with oil, spices (mostly turmeric), and dill, served with a plate of greens to toss into the pan. It is delicious, although on one occasion my dining companion expressed surprise — he was expecting far more for his US$5. Quality rather than quantity seems to be their policy. Cha Ca Vong is generally washed down with copious quantities of cold beer.

Apart from all the delicious specialties, the basis of Vietnamese and Indochinese food is the fermented sauce known in Vietnam as *nuoc nam* — a pungent and flavorsome mix of certain fish varieties fermented with salt for periods of up to 12 months. It becomes quite delicious when mixed with chilies, lime juice, sugar and garlic, giving new meaning to simple rice (*com*) dishes. Another delicious dip comes with freshly steamed crab: a simple mixture of black pepper and salt (more often monosodium glutamate) and lime juice which gives a tangy bite to any seafood.

LAO CUISINE

Across the border in Laos, the cuisine shares similarities with Thai food, especially in the liberal use of fresh ingredients, lime juice, plentiful hot chilies, cilantro (coriander), and fermented fish concoctions.

Unlike the Vietnamese love for noodles, the Lao staple is the almost indigestible sticky rice (*khao niaw*) which is rolled into ping-pong-sized balls and consumed in terrifyingly vast quantities along with small portions taken from side dishes. A generous mixture of spices go into Lao cooking, including lime juice, lemongrass, chilies, galangal root (a cousin to the ginger plant), marjoram, basil, cilantro (coriander), ginger, and tamarind.

One of the best dishes in the fairly limited Lao repertoire is *laap* (pronounced larp), a delicious fresh mix of a minced meat (beef, *sin ngua*, or chicken, *kai*), spices, lime juice, garlic,

OPPOSITE: Small sweet raspberry-like fruits are eaten with salt in Da Lat. BELOW: A woman selling bread outside her house in Luang Prabang.

vegetables (*phak*) and chilies, served with sticky rice in individual bamboo containers and accompanied by a salad of lettuce, mint and other foliage.

Another delicious staple reminiscent of Thai food is *tam som*, a spicy salad of sliced green papaya, chilies, lime juice, garlic, and condiments, all pounded together and served with sticky rice.

Luang Prabang has its own special cuisine which includes watercress salads and a kind of fried river-seaweed wafer decorated with sesame seeds, known as *khai paen* — an acquired taste, but certainly interesting. Chinese- and Vietnamese-influenced foods are also popular and easy to find in Laos. The noodle soup known as *pho*, Vietnamese-style fried spring rolls known as *yaw jeun*, and Thai-style marinated and grilled chicken are also popular and very edible.

Perhaps the most evil of all Lao condiments is a coarse, fermented fish paste called *paa daek*, that is found on restaurant tables. Apart from being a health risk to the uninitiated, it smells like the scrapings of a long-dead corpse, and is definitely not recommended.

Another Lao specialty, although not quite a condiment, is the fiery *Lao Lao*, a distilled rice liquor which helps to pass a day quite merrily. Once, stranded by a broken down bus in the middle of nowhere, fortunately close to a small farmhouse and stall, I watched passengers disembark to gratefully down thimblefuls of Lao Lao, becoming all the merrier as the hours dwindled by. When the time came to leave, some of the passengers were so relaxed they stayed behind, quite happy to continue their journey another day. That's the sort of place Laos is.

KHMER CUISINE

Possibly Khmer food once enjoyed a degree of sophistication, but since the depredations of the Khmer Rouge, the cuisine has never recovered its grandeur, except perhaps in a few specialty restaurants that feed Khmer specialties to the tourists (try the specialty restaurant in the Grand Hotel d'Angkor in Siem Reap). Generally, the cuisine is quite a basic, although satisfying, blend of rice served with fish and vegetable dishes. Curries and soups add variety. Luckily there are plenty of restaurants in the capital and around Siem Reap catering to decadent foreigners and their sophisticated palates.

Special Interests

ARCHITECTURE

Intact, sprawling, colonial villas, neoclassical French frippery, low sweeping Lao temple roofs, gold-lacquered pillars, handmade tiles molded over the thighs of young virgins, dragons, half-man half-bird statues, beautifully detailed stucco temple frescoes, ancient Khmer temple cities, the thatched roof and woven bamboo walls of highland tribal homes, monumental Soviet-style public buildings, and neo-Stalinist concrete monstrosities are just some of the architectural quirks to be discovered across Indochina.

The Chinese-influenced Vietnamese imperial architecture can best be seen in Hue in the tombs of the Nguyen Dynasty, scattered across the countryside. In Hanoi the stately lines of the Vietnamese Temple of Literature mingles with classical French architecture, with many streets remaining in almost their original condition. Fortunately, much of Hanoi's newer development has been engineered outside the city center, thereby retaining a degree of architectural integrity rare in an Asian city. Grand French neoclassical buildings can be found lining the riverfront of the old port city of Hai Phong, while Da Lat has been nicknamed Little France, so numerous are its colonial buildings. Vientiane, Phnom Penh and even to a lesser degree Saigon (its older buildings and temples are still there but well hidden behind all the new high-rises) all have fine colonial buildings to be proud of.

The ancient Chinese/Japanese trading port of Vietnam's Hoi An is a great architectural find where narrow streets are lined with over 500 nineteenth-century and earlier Chinese and Japanese buildings. Deep-set Chinese houses (to a depth of at least 35 m, or 115 ft) stand next to pagodas and temples. Ancient clan houses have been converted to bars while others fulfil their original function, to be discovered along with shrines, community houses, and the Japanese covered bridge which once connected the Japanese quarter, with its Japanese-style houses and shops, to the Chinese quarter. Although many of the old buildings are being converted into galleries and bars, the original framework remains. The eager-to-please **Hoi An Tourism Company** ((51) 861 373 or (51) 861 332 FAX (51) 861 636, at 6 Tran Hung Dao, sells tickets that provide entrance to some of the best old houses.

Cambodia's ancient Khmer wats and temple cities, namely Angkor Wat, provide plenty of food for thought in the structure and layout of the major complexes, to say nothing of the superb stone carved decorations. Architecturally speaking, Indochina is a gold mine.

PAINTING

Since the advent of *Doi Moi* and Vietnam's lightening of official Communist policy, Vietnam has seen an

OPPOSITE: A stall on Nha Trang beach is crowded with snacks and beer. ABOVE: The well-tended gates of Phnom Penh's Royal Palace signify order and riches.

explosion of talent in the art world. Painters who went underground or turned to propaganda art during the dark years were suddenly free to paint again, expressing in vivid color and form all the images that were forced to remain dormant for so long. Vietnam has a long tradition of arts and painting. The French colonists had started the School of Arts of the Far East early in the twentieth century, giving young Vietnamese students a thorough grounding in expressionism and the French classicists. This combined with a long tradition of Vietnamese folk arts leads to exciting work that is being recognized internationally. Needless to say there is plenty of fluff mixed in with the true

talent, but an exploration of the better galleries in Hanoi and Saigon will yield satisfying results. Check with the locally printed guides for current exhibitions in both cities to see exciting new works. A Hanoi based art consultant, Suzanne Lecht ((4) 862 3184 FAX (4) 862 3185 E-MAIL suzlecht@netnam.org.vn, sometimes takes visitors around to meet both established and aspiring Hanoi artists, and will help guide buyers through the maze of artworks to buy worthwhile works, for a small commission.

HANDLOOM TEXTILES

The absorbing world of handloomed textiles is particularly important in Laos where every mature woman has an intimate appreciation of weaving and is expected to wear a handloom skirt (*sinh*) as part of her daily wardrobe, and where it would be unthinkable to enter a government office wearing anything else. Even in the streets of Vientiane, it is very unusual to see women wearing trousers or factory-made dresses.

A growing number of collectors are taking an interest in the distinct regional patterns and styles of Lao textiles. Most of the cloth is woven from silk and a visit to a market will reveal skeins of creamy-colored undyed silks, as well as richly-hued pre-dyed skeins hanging from the walls.

Laos and Cambodia are both known for the superb weft silk *ikats,* where the weft threads are dyed with the patterns before weaving. When five or six colors are involved, natural vegetable dyes are used and the pattern is small and complex, it can take months just to prepare the yarns for weaving. Cambodian and southern Lao *ikats* are probably the finest and most detailed in the world.

Currently, with the publicity given over the last few years, the prices have escalated rapidly, now nearing international prices. Those willing to spend time hunting around can still pick up some marvelous bargains, although even the bargains will cost a lot more than a few pennies.

An excellent book to help through the complexities of Lao handlooms is Mary Connor's *Lao Textiles and Traditions* published by Oxford University Press in 1996.

TRIBAL CULTURES
Tribal cultures exist all throughout Indochina, many living a life hardly different from their ancestors. Those wishing to explore are requested to tread gently; these cultures are fragile, many still working mainly with a barter economy. Camera-flashing, money-spending tourists can quickly upset ways of life that have been operating perfectly well for centuries. Some of the more

interesting tribal areas are in the far north of Laos and Vietnam, close to the Chinese borders — Muang Sing, Oudomxai, Phong Saly, Sa Pa, Bac Ha, Dien Bien Phu. Intrepid Tours does a recommended 16-day "Hill Tribes of Northern Laos" trip that guarantees visits to interesting and little-visited villages with two days of trekking (see TAKING A TOUR, below). They also do a seven-day "Trekking in Sa Pa" tour visiting Vietnamese minority villages.

Taking a Tour

There are now so many competent tour companies operating in Indochina that anyone planning an organized trip will be spoiled for choice. Tour companies can be particularly useful for short one- and two-day tours out from a city such as Hanoi, and for tours to more remote parts of the country such as the northern highland areas.

OPPOSITE: Hand-woven baskets in Phnom Penh market sell for cents. ABOVE: A man poles his sampan down a placid river in Vietnam's central highlands.

For visiting established tourist centers such as Laos's Luang Prabang, Saigon or Hanoi, a tour is really unnecessary. A day-long city tour provides a good orientation and overview of the main sights, after which you are just as well off taking things at your own pace. Tour companies prove their usefulness for specialty tours and there are many interesting tours on offer in addition to the exhausting, "see Indochina in a week" type that please no one.

The ultra-expensive and efficient **Diethelm Travel** ((21) 213 833 or (21) 215 920 FAX (21) 216 294 or (21) 217 151 E-MAIL dtllvte@pan-laos.net.la, based in Bangkok with an office at Setthathirat Road, Nam Phou Square, Vientiane, Laos, specializes in Indochina, offering a range of tours to suit to all tastes, including "soft adventure tours" with easy trekking in northern Laos, or sedate rafting on the Nam Xong River in Vang Vieng, 150 km (93 miles) north of Vientiane.

Two highly recommended tour companies in Vietnam are the small, private **Especen Travel Agency** ((4) 826 6856 or (4) 826 1071 FAX (4) 826 9612, based in Hanoi at 79E Hang Trong Street; and the efficient, state-owned **Saigon Tourist** ((8) 230 102 or (8) 230 100 FAX (8) 224 987 or (8) 225 516, at 49 Le Than Tho Street in Saigon. Both offer a wealth of well-run tours which include tailor made, client-specific

tours as well as targeted special tours. They offer Environment and Rural Life Tours, Business Tours, Camping Tours, Trekking Tours, War Veteran Tours to major battle sites, Rail Tours, a Cham Civilization Tour of Central Vietnam, Trans-Vietnam Tours, and extension tours to both Laos and Cambodia.

A Bicycle Vietnam Adventure is available from a couple of international companies: the Australian organization **Intrepid Travel** ((61 3) 9416 2655 FAX (61 3) 9419 4426, PO Box 2781 MDC, Fitzroy, Victoria 3065, Australia, and **British Exodus Adventure** ((44 181) 675 5550 FAX (44 181) 673 0779 E-MAIL sales@exodustravels.co.uk, whose head office is at 9 Weir Street, London

SW12 OLT (see SPORTING SPREE, page 32). Intrepid Travel also offers several other tours including an excellent tribal tour of northern Laos (see SPECIAL INTERESTS, above).

Sea canoeing tours to Ha Long Bay are available from **Sea Canoe International** ((66-76) 212 252 FAX (66-76) 212 172 E-MAIL actioninfo@seacanoe.com WEB SITE http://seacanoe.com, which is based in Phuket, Thailand (see SPORTING SPREE, page 32).

In Laos, **Inter-Lao Tourism** ((21) 214 832 FAX (21) 216 306, on the corner of Setthathirat and Pang Kham Roads, and **Lao Travel Service** ((21) 216 603 FAX (21) 216 150 E-MAIL lts@pan-laos.net.la, 8/3 Lan Xang Road, offer everyday tours,

trekking tours around Luang Prabang, northern Laos trekking tours visiting various hill tribes, and tours exploring southern Laos.

There are also several backpacker companies who, while catering mainly to the backpacker market, offer reasonable service and great value for the travel dollar, especially in Vietnam. The best of the cheapies (for day and short trips around the north of Hanoi) is **Green Bamboo** ((4) 826 8752 FAX (4) 826 9179, 42 Nha Chung Street.

OPPOSITE: Hmong boy stands near the Cloudy Bridge in Sa Pa. ABOVE: Hmong girls walk their way to Sa Pa town.

Welcome
to
Indochina

ON ONE OF MY MOST RECENT VISITS TO VIETNAM, I was traveling from Hanoi to Hai Phong when a flash of light — something as swift and elusive as a fish rising and diving in a lake — caught my eye in the rice paddies that flank the road just beyond the capital's urban limits. I stopped the car and took a closer look, and watched a rural ritual that's become something of a dream for Asia travelers in this era of burgeoning industrial economies.

On a grassy dike alongside an irrigation stream stood two gnarled old women, each day rituals of rural life — traditional farming techniques that have given way to mechanization in many other areas of Asia — that are almost a kind of ballet, a folk art. I watched for different irrigation techniques, in particular, and found that they varied from region to region. In other areas of the north, the bamboo basket became a heavier wooden scoop hung from a bamboo frame and rocked back and forth with the same endless rhythm, like a baby's cradle, to hurl the water from one paddy to the next. On the road between Da Nang and Hue in the northern

accompanied by a younger woman who was obviously their daughter, or daughter-in-law. Working as two teams, they faced each other on the dike, bending and swaying in a sinuous, almost hypnotic rhythm — each couple pulling on woven bamboo ropes attached to a wide-mouthed bamboo basket. As they swayed forward, the basket sank into the irrigation canal. As they pulled back it lifted, full of water, and with an almost imperceptible flick of the wrists they made it flip and dash its contents into an adjacent rice field. It was the water, splashing in the morning sunlight, that had caught my attention on the road.

This was something I was to see time and time again throughout Vietnam, the every-

central region of what used to be South Vietnam, I came across two old people — husband and wife — sitting side-by-side on a bamboo contraption in their wide-brimmed conical bamboo hats, and pedaling a wooden machine with a paddle-wheel as it scooped the water into a small field of rice seedlings. From a distance they looked like lovers sitting together in the golden, late afternoon sun.

When I approached them and talked with them, they had the same friendly, guileless charm that I was to find in the countryside right across Indochina — a charm that's all the more striking and enjoyable when you

LEFT: Tending the rice fields, Vietnam's Mekong Delta. ABOVE: Beach vendor, Quy Nhon, Vietnam.

consider not just the backbreaking reality of peasant life but the long shadow of conflict and suffering that lies behind the color and exoticism of this entire region. And it's this juxtaposition of charm, antiquity, and epic struggle that makes this area the most culturally fascinating new destination in Asia.

Indochina, one of the most beautiful and most cultured regions of Asia, has a turbulent and often violent history which extends more than 2,000 years into its past. If we consider just these two millennia alone, the three countries that make up this beleaguered region — Vietnam, Laos, and Cambodia — have been engaged in almost continuous internecine wars and major struggles to survive in the shadow of their powerful neighbors, China, Siam (Thailand), and Myanmar (Burma).

Centuries of conflict have given them a tenacity which has been most dramatically evident in their heroic, if devastating, contemporary struggles — against the French, who colonized Indochina from the late 1800s, and against the Americans, who decided in the early 1960s that the toppling dominoes of Communist expansion would be halted in the jungles, mountains and fecund rice plains of this tragic region. The brilliant victories scored against both these Western powers have only heightened the sense of tragedy. Military triumph has won Indochina the political and cultural independence that its three territories have been seeking for centuries, but at a tremendous cost: in each case, victory has placed them under hard-line revolutionary socialist rule, resulting in nearly two decades of social purges, isolation, austerity, and virtual economic collapse.

Only now are these unique cultures, overshadowed for so many years by conflict, emerging from their Communist straitjacket with all the color and exoticism of cultural renewal. Now, as these three countries re-open their battered economies to the rest of the world, the future beckons. Laos is upgrading policies, roads, and facilities across the country in readiness for Visit Laos Year 1999. Previously off-limit areas are being opened, land mines by the thousands are

Mountains of rice varieties sell in a market in Vietnam's Mekong Delta.

being cleared from troubled areas like Xieng Khuang Province, in readiness for the hordes of tourists expected in the years to come.

Although much of Indochina remains "backward" by the standards of developed countries, this very backwardness makes the region an adventure for travelers. Industrialization has not yet had a chance to annihilate traditional cultures, so we enjoy the privilege of visiting a region where Asia is still Asia — lands of gracious and devout Buddhists, of golden-spired temples and exotic customs, of cone hats and buffalo

plows in verdant rice paddies, seemingly undeterred by all the madness that has surrounded them.

While tour groups will experience the usual kid-glove treatment, and independent travelers will certainly experience the odd difficulty or, at the very least, some discomfort, the sights, sounds and experiences that await discovery in this exciting region more than compensate for any minor discomfort.

A great pilgrimage is already under way — perhaps greater than the initial rush to visit "forbidden" China when it first reopened its

ABOVE: The benign visage of a god king, Wat Thom Angkor. OPPOSITE: Hill-tribe village TOP on the Ha Giang River, Vietnam. Sunrise at Dien Bang BOTTOM near Da Nang.

doors in 1979. The word is out and quiet unassuming Laos is being "discovered" by thousands of visitors who are leaving the country charmed by what they have experienced. Cambodia and Vietnam, which have been attracting visitors for years, offer fewer surprises but much to experience.

Throughout much of Indochina, the still-present French colonial architecture is being restored, with a growing civic pride in old architecture hampering any trends in rebuilding — the beautifully-restored old Municipal Theater, more commonly known as the Opera House, in Hanoi; the elegant central Post Office in Saigon, with its main hall reminiscent of a cavernous Victorian railway station; and a fading but particularly stately mansion which is now the home of the Laos Revolutionary Museum in Vientiane are just a few examples. The weather-worn but picturesque terraces and old public buildings of the former royal capital of Laos, Luang Prabang, nestles among the golden spires of Buddhist wats as though a sleepy French provincial town has been lifted up and set down in Shangri-La.

Traveling through Indochina today, it's almost as if the relics of colonialism have been fiercely preserved — though it's probably more realistic to consider that, within the alternating cycle of war and austerity that's gripped the region right up to this day, there's been little time or money to pull the old French infrastructure down and build a new one in its place.

What's left of the Vietnam War has nothing of the French charm — rusted, stripped helicopter gunships, some left where they crashed in the countryside, others mounted outside the war museums; a crudely-sculpted stone plaque alongside Truc Bach Lake in Hanoi commemorating the shooting down of a United States aircraft in 1974; as we prepare to enter the new millennium, many such sights remain to remind the world how high-tech military power failed against an agrarian bamboo culture. From the air, you can still make out the craters left by countless B-52 bombing raids over the Plain of Jars in Laos or over Da Nang in Central Vietnam. Flying into key airports throughout the region, you can see rolling clusters of concrete shelters and hangars which once housed

American warplanes. At the top of the hill called Phou Xi in Luang Prabang, right next to That Chom Si, you'll come across an old Russian antiaircraft gun that was presumably installed after the Communist Pathet Lao victory. Then there are the war cripples and amputees, thousands of them, especially in Vietnam and Cambodia. And Cambodia, although Pol Pot is dead and gone and the Khmer Rouge currently stand on their last legs, is yet again plunged into anarchy and lawlessness, and the next chapter in this long and violent story waits to be written.

For many veterans in this new tourist pilgrimage, especially to Vietnam, Indochina is an emotional experience, a coming to terms, a healing of wounds; and it's reflected in the desire to make amends that's drawn some of them back — groups of vets building a clinic outside Da Nang, a school near Phonsawan, the new tin-roofed "capital" of Xieng Khuang Province in Laos, home of the Plain of Jars. The old capital, Xiang Khuang, was virtually destroyed by bombing in the war. But for all visitors, whether they were part of the war or not, there is something about Indochina which cannot be found anywhere else in the world — the juxtaposition of horror and beauty, the region that was engulfed on the one hand in one of the most

savage, destructive wars in history and is now, in peacetime, unveiling a mystique, exoticism and charm that were virtually lost in the clamor of television news reports and newspaper headlines in the past.

Right across the cultural spectrum, Indochina is reintroducing the world to its rich history and culture — magnificent Hindu and Buddhist temples and relics like the fabled Angkor Wat in Cambodia, the Cham ruins in Vietnam and the region's oldest Buddhist landmark, Wat Phu in Laos; evocative traditional dance and, particularly in Vietnam, unique musical instruments; dramatic natural attractions like the offshore karst formations of Vietnam's Ha Long Bay and the superb beaches that run the length of Vietnam's long coastline. Sites like Laos's enigmatic Plain of Jars, which have been off-limits for decades, are now accessible. Some of the most beautiful countryside in Asia, can be seen in Vietnam's Central Highlands and the vast Mekong Delta region, and in the rugged mountains and plateaus of Laos. Rural life where traditional farming techniques are still in practice, bring you marvelous images of an Asia that's fast disappearing in Indochina's more economically developed neighbors.

The people themselves are friendly, charming and devoutly Buddhist, and perhaps as part of their Buddhist or Asian heritage spend no time wallowing over past tragedy. And yet, Indochina still stands in something of a time-warp, its architecture and infrastructure frozen in time with nothing essential changed. For newcomers, Indochina offers the opportunity to explore vibrant, little-known cultures where time has virtually stood still. It won't last. Thailand too, was charming and gracious just twenty years ago, and for the last glimpses of a vanishing era, now is the perfect time to visit Indochina before "development" and industrialization cause too many changes in these, the last of Asia's pre-industrial economies.

ABOVE: Cham ruins at Phan Thiet, Vietnam.
RIGHT: The hills and valleys of Phan Rang, Vietnam.

Welcome to Indochina

Indochina
and Its
People

ARCHAEOLOGISTS HAVE FOUND EVIDENCE of human settlement in Indochina, mainly in what is now northern Vietnam, as early as 500,000 years ago. But it wasn't until the thirteenth century BC, during the Bronze Age, that anything approaching sophisticated tribal life appeared in the coastal plains of Vietnam, beyond the hardships of the vast jungle-clad mountain chains that march up from the northern fringe of the Mekong Delta into landlocked Laos.

THE ANCIENT HINDU KINGDOMS

By the time of the birth of Christ, the first of two great cultural influences was beginning to give shape to Indochina. The Hindu influence, which was to sweep the region from India to Indonesia, established the kingdom of Funan in what is now southern Vietnam and eastern Cambodia. From the first to sixth centuries AD, absorbed into the vast trading empire that India had created, Funan literally put the Indochina peninsula on the map — enjoying prestigious and far-flung trading contacts with India, China, Indonesia, Persia and the Mediterranean. In one of those discoveries that makes any historian's heart leap, archaeologists have unearthed a gold Roman medallion in Vietnam's Kien Giang Province, where Funan's capital was located, dated AD 152 and inscribed with the bust of Emperor Antoninus Pius, the successor of Hadrian. This early internationalism, and the colonial era that came much later, has a significant bearing on Vietnam's economic renaissance today: the south, particularly, has a long tradition of trade and cultural exchange with the rest of the world.

While Funan was in its heyday, another Indianized domain, the Hindu kingdom of Champa, arose to the north in the coastal area around what is now Da Nang. While Funan has left an important commercial legacy, Champa's legacy to Vietnam remains its most illustrious cultural relics — the Cham ruins, remains of 15 temple towers, at My Son near Da Nang, and a collection of stone images and sculptures in the city's Cham Museum.

By the eighth century, the Champa Kingdom dominated much of what is today central Vietnam, extending its rule as far south as Phan Rang. But it was constantly at war

with the southward-migrating Viets to the north, who in turn were engaged in a long-running struggle for survival against imperial China.

THE CHINESE IMPACT

The Vietnamese themselves, though, originate from Eastern China in an area now covered by the provinces of Jiangsu, Zhejiang and Fujian. Until the king of the state of Qin unified China in the third century BC, China was a collection of independent kingdoms, two of which were called Yueh and Min-Yueh. "Yueh" is the Chinese word for "Viet," the ideogram comprising the radical for "roam" with the word "bandit." The two kingdoms were known for their irreverence

and failure to observe the norms of traditional propriety — Confucius grumbled in the fifth century BC about their lascivious women and vulgar music.

"Yueh-nan" or "Vietnam" means the Southern Viet, and south they went, spurred on by the efforts of the king of Qin (who had become Emperor of China) to secure their obedience. They migrated into Champa, Funan and Chenla, conquering and absorbing the local population as they went. Many Cambodians and Laos would argue that the trend has not stopped. In spite of their resistance to Chinese domination, Vietnam's culture still owes more to Chinese traditions than those of either Cambodia or Laos. The Chinese had conquered the Red River Delta in northern Vietnam in the second century BC,

triggering a fierce resistance among the Vietnamese that still underscores relations between the two countries today. While relations between Hanoi and Beijing are now normalized, with both clinging to the collapsing pedestal of socialism, they fought a short but bitter border war as recently as 1979 and have yet to settle their rival territorial claims to offshore oil deposits of the Spratley Islands in the South China Sea.

The first most memorable battle against Chinese central rule was fought in the year AD 40 by the Trung Sisters, two high-born Viets who led a rebellion that reestablished sovereignty over the Red River Delta. Three

Statues of mandarins guard the Khai Dinh imperial tomb in Hue, Vietnam.

years later the Chinese counterattacked, grabbed the region back, and the two vanquished queens committed suicide; but they've gone down in history as Vietnam's most revered heroines, and even today you'll find key streets named Hai Ba Trung in Hanoi, Saigon and other major cities.

For more than 800 years, right up to the tenth century, the Chinese maintained and strengthened their grip on today's northern Vietnam, naming it Annam ("pacified south"). There were various Vietnamese rebellions during this period, but it took the collapse of the Tang dynasty and China's preoccupation with chaos and power-struggles within other parts of its realm, to give the Vietnamese the advantage in the field. In 938 the Chinese were finally routed in a revolt led by Ngo Quyen. A thousand years of Chinese imperial rule came to an end, and Ngo established the first of 12 Dynastys that were to maintain Vietnamese sovereignty and independence, despite various internal conflicts, until it fell under French domination in the mid-1800s.

But it must be remembered that for much of this time, what we now regard as Vietnamese sovereignty applied only to the northern region. The state of Champa continued to dominate what was to become central Vietnam and part of the south right up until the eighteenth century, creating a cultural division between north and south that has been most vividly emphasized in modern times by the 1954 partition of the country and the Vietnam War, and is even quite evident today. As for Funan, we'll see that it simply disappeared into the maw of the Khmer empire, based in Cambodia, which controlled the Mekong Delta until this vast, rich rice bowl was brought under Vietnamese suzerainty in the seventeenth century.

Nor did the landmark battle of 938 see the last of northern invaders. In the thirteenth century the Mongol Yuan Dynasty, the heirs of Kublai Khan's savage conquest, invaded northern Vietnam twice in a bid to get their hands on Champa. They were soundly defeated on both counts, and two more great

This well-preserved Cham tower dates from the sixth century, at Po Nagar, near Nha Trang, southern Vietnam.

warriors joined the ranks of Vietnam's resistance heroes — Tran Hung Dao, whose statue stands today, high on a pedestal in Saigon's "Hero Square" and Emperor Ly Thai Tho, better known as Lei Loi, whose name has joined that of the Trung Sisters on several main streets throughout Vietnam.

ARRIVAL OF THE MOUNTAIN PEOPLE

While Chinese pressure to keep its empire intact can be said to have welded Vietnam-

While the Thais settled right through northern Laos and Thailand, they clung together in separate tribal groups, each with its own leader. But in the mid-thirteenth century the clans in northern Thailand organized themselves in rebellion against the Khmers, and in doing so established the kingdom of Sukhothai. In a series of pacts with Thai warlords in Chiang Mai and Phayao, Sukhothai rule was able to extend right across the Mekong to include Wieng Chan (City of the Moon), whose name the French later romanized into Vientiane.

ese nationhood, it was the crucible from which neighboring Laos sprang. It's known that there was a human presence in this largely mountainous wilderness some 10,000 years ago. But around the eighth century, the first of two major migrations by Thai-Kadai tribes from southern China began trickling through the steep hills, finally reaching the more benign Mekong River Valley and either settling there or continuing on into what is now northeast Thailand. Five hundred years later another migration began, this one fleeing Mongol troops sent into China's Guangxi and Yunnan Provinces to bring troublesome non-Han populations to heel.

Thwarted in northern Thailand, the Khmers now gave their backing to the man who could well be called the father of Laos nationhood — Chao Fa Ngum, a warlord from Muang Sawa (later known as Luang Prabang). Fa Ngum seized Wieng Chan from the Thais and then marched on into Thailand itself; and in 1353 he established the kingdom of Lan Xang (Ten Thousand Elephants) embracing the Khorat Plateau of northeast Thailand and much of what is the state of Laos today. Fa Ngum didn't stop there — he pushed Lan Xang's borders eastward to the Annamite Mountains of Vietnam and even threatened the kingdom of Champa, which unnerved his ministers so much that they deposed him and booted him into exile.

Fa Ngum's successors consolidated what was now a powerful kingdom and the embryo of Lao nationhood. But when one king, Setthathirat, went missing in 1571 after a military foray into Cambodia — thought to have been slain by unpacified hill tribes in southern Laos — Lan Xang fell into 60 years of chaos during which the Burmese, from the north, took the opportunity to pick bits of the kingdom off for themselves. Finally, a new iron leader, King Suliya Vongsa, arose out of the turmoil in 1637 and, in the ensuing six

BIRTH OF THE KHMER EMPIRE

At this stage of Indochina's history, with Vietnam comprising two relatively powerful independent states and Laos torn apart by internal rivalry, it's time to look at the third key element of the saga: Cambodia. From the first to sixth centuries, much of what is now Cambodia was part of the Hindu kingdom of Funan, and it shared the region's brisk trade and cultural intercourse with the outside world. But in the middle of the sixth

decades — the longest reign of any Lao monarch — pulled the warring Lao factions back together, reestablished firm centralized rule, and made Lan Xang more powerful than it had ever been before. This was the kingdom's golden era, its Elizabethan age. The trouble was, King Vonsa died in 1694 with no son to succeed him, and this first Lao Kingdom was split into three warring kingdoms — Lan Xang, Xieng Kuang and Champassak. More than that, the Laos found themselves caught in a deadly pincer, with the Thais of Siam on one side of them and the Vietnamese on the other — each of them intent on claiming this strategic mountain domain as a vassal state.

century, the reign of the Khmers began — an age of explosive expansion characterized by conquest and tremendous cultural development.

In the middle of the sixth century a tribal force called the Kambujas from the middle Mekong region established a new kingdom called Chenla. This state then turned its attention east, to the rich Mekong Delta area, and began to absorb Funan. Two centuries later, Chenla broke up into rival northern and southern kingdoms. Civil wars broke out in the south as rival leaders fought for power, weakening the region so much that it

OPPOSITE: A Van Kiew tribal village in the Cam Lo Valley, central Vietnam. ABOVE: Young fisherwoman in Da Nang.

attracted invasion and annexation by Hindu forces from Java's powerful Sailendra Kingdom.

It took two attempts to sever Cambodia from Javanese rule, which was all-powerful in the region at the time. The first try by a Khmer prince ended in short order when the Sailendra ruler mounted an expedition to Cambodia and beheaded him. The second effort was made in 802 by Prince Jayavarman II, who declared himself independent of Java from his mountain fastness at Phnom Kulen, the first capital of Cambodia. The Javanese were unable to bring the Cambodians to heel and in ignominy watched the rapid growth of a great imperial power under Jayavarman. More importantly, he established a dynasty in which successive kings built upon the firm centralized administration that he'd set up. The result was the Khmer Empire — which by the twelfth century had brought a large part of Vietnam, Laos and Thailand under its control and had even subjugated the kingdom of Champa.

This move against Champa was rash and ill-fated: Cham armies struck back in 1177, destroyed Angkor and very nearly wiped out the Khmer empire with it. Had a Khmer strongman not stepped into the breach at this point, Cambodia would not possess one of the great cultural wonders of the world today. The empire's savior was King Jayavarman VII, who ascended to the throne in 1181 and immediately began restoring Khmer stability, sovereignty and power. But his fame rests in the new symbol of Khmer power that he built — Angkor Thom. Many people today think of Angkor as Angkor Wat, built by Suryavarman II (1112–1152) — a single, particularly spectacular temple ruin, albeit one of the most important on earth. But what Jayavarman VII built was a huge new capital, Angkor Thom, with a population of more than one million, an intricate network of canals, dams and irrigation systems and Angkor's most celebrated monuments, the awesome Bayon Temple, the Baphuon and the Terrace of Elephants.

An important point worth noting is that in spite of the greatness of the Cambodian Empire, it was fatally flawed with a mysterious problem that shaped its policies and sowed the seeds of the monstrous brutality

which reached its apogee under the Khmer Rouge. Its people seemed incapable of producing sufficient progeny to maintain the population or provide adequate labor for the great engineering works being undertaken at the time. The consequence was that the only way in which the Khmer kingdom could maintain its population level and get enough laborers was to launch an endless succession of wars on its neighbors, not in order to acquire territory but to obtain slaves who were atrociously treated. Far from being a new phenomenon, the Khmer Rouge are the inheritors of the darkest side of an ancient and glorious history (The only contemporary account of life in Cambodia in those days survives as notes made by an "ambassador" of the Mongol court in Khanbaliq — later

called Beijing—who was in fact a spy named Chou Ta-kuan. He was sent to assess the possibility of Mongol conquest, and his advice was "You haven't a hope!" His notes make hair-raising reading).

Reinvigorated by all this grandeur, the Khmers embarked on another period of conquest and avenged their humiliation by the Chams by driving deep into Champa and ultimately destroying this proud central Vietnamese state. But then, as with the Lan Xang empire of Laos, Khmer power waned with the death of its iron man. With the loss of Jayavarman VII, Angkor gradually fell into decline and the Khmer state found itself weakened in the face of an onslaught by the rising new power of Southeast Asia — the Thais.

By the end of the sixteenth century, both Cambodia and Laos were virtually fighting for their lives, threatened and squeezed by the Thais and the Vietnamese. Both states remain trapped to some extent in the same vice today, with Vietnam and Thailand competing for cultural, political and economic dominance of them. In Laos, the death of King Suliya Vongsa in 1694 and the collapse of the Lan Xang state eventually led to Vietnamese control of Wieng Chan (Vientiane) and the Middle Mekong region and Thai control of Champassak to the south, a rich agricultural plain close to modern-day Pakse. It was when the Thais began confronting the Vietnamese in Wieng Chan, exacting tribute

Angkor Wat temple ruins — the sacred city had a population of over one million in its heyday.

alongside their rivals, that Laos's subservience ended.

The area's warlord, Prince Anou, boldly declared war on the Thais, with disastrous results: the Siamese destroyed Wieng Chan, then marched on Luang Prabang and Champassak, and by the closing years of the eighteenth century had occupied and virtually depopulated most of the country, forcibly resettling the Laos in northeast Thailand.

In Cambodia, Siamese attacks began in the thirteenth century after the death of Jayavarman VII, but it took the Thais more

than 200 years to secure the prize they were after — the vast treasure house and seat of Khmer power, Angkor. It was in the mid-fifteenth century that they finally overran the temple city, forcing the Khmers to move their capital to the vicinity of what is now Phnom Penh. The Khmers fought back, and at one stage managed to push their forces to the Siamese capital of Ayutthaya; but in 1594 the Thais conquered Phnom Penh, and the great age of Khmer power and prestige was finally brought to an end.

An equally significant aspect of the Khmer collapse is how it introduced a completely new political and military element into the Indochina arena, and one which would have a profound effect on the region

in the future. Pressed by the Thais, and with Phnom Penh about to fall, the Cambodians became the first Indochinese to call upon the Western powers for help. The request went out to the Portuguese and Spanish, the first Europeans to explore Asia; but when a Spanish expedition from Manila finally came to the rescue it was too late — Phnom Penh had already been conquered by the Thais.

Instead of turning on the Thais, or returning to Manila, the Spanish decided to stick around. They then became so deeply embroiled in Thai-Cambodian intrigues that the Khmers eventually had to do a complete turnabout and enlist Thai help to get rid of them. In 1599 the entire Spanish garrison in Phnom Penh was massacred, and a Thai puppet monarch put in power. From that point on the once-powerful Khmer state was ruled by a succession of beleaguered monarchs who were under such intense pressure from rivals that they had to seek Thai or Vietnamese help to stay in power.

But these were costly alliances: the Vietnamese began their takeover of the southern Mekong Delta region — once a Khmer domain — in return for their assistance, and the Thais grabbed control of Battambang and Siem Reap Provinces. Not only that, they grabbed the Khmer royal family, too, actually crowning one Khmer monarch in their own capital Bangkok and then installing him in power at Udong, a temple city close to Phnom Penh. Indeed, the Khmer state — and modern Cambodia — might not have prevailed at all, had it not been for the timely, if infamous, intervention of yet another foreign power which has had the most abiding contemporary influence on Indochina — the French.

THE FRENCH

The French colonial experience in Indochina followed the rather bumbling tradition of European colonialism across the globe — exploration, followed by trade, followed by missionaries, then culminating in military action to protect the merchants and evangelists from the inevitable cultural backlash. And this is almost exactly how the French stumbled into Vietnam.

Although Portuguese, Dutch and French traders and missionaries had been active in Vietnam since the sixteenth century, the French were the first to establish a military presence there. We've read how a series of 12 Dynastys, beginning in 938, were able to maintain sovereignty and a measure of stability in northern Vietnam, with the kingdom of Champa flourishing in much the same state in the central-south. In 1765, the Tay Son rebellion, led by three brothers from a wealthy trading family and aimed at corruption and poor administration, put an end to

Vietnam's military hall of fame and established yet another tradition of heroic street names.

But while Nguyen Hue was saving the north, a prince named Nguyen Anh, usurped by the Tay Son rebellion, was plotting revenge in the south. After getting nowhere with the Thais, Nguyen Anh made elaborate overtures to the French for military assistance, even sending his four-year-old son to the court of King Louis XVI as a gesture of good faith. The support that he got was tacit, but enough to give him the edge over the Tay

the status quo. Within eight years, the Tay Son rebels had overrun southern Vietnam, killing 10,000 Chinese residents of Cholon in what is now Saigon in the process; then they went on and conquered the north.

At this point, the route to French dominance of Vietnam gets a bit complicated, but I can assure you that it all becomes clear in the end. First of all, the emperor installed in the north by the Tay Son rebels turned out to be a weakling who called in 200,000 Chinese troops to help him maintain power. This led to one of the rebel brothers, Nguyen Hue, proclaiming himself emperor and raising an army to kick the Chinese out — which he did in 1789, routing them in a battle near Hanoi which earned him a revered place in

Son. With the help of two warships and 400 French mercenaries, Nguyen Anh crushed the Tay Son, captured Hanoi, proclaimed himself Emperor Gia Long and set about rebuilding the war-ravaged country. More significantly, he united Vietnam — north and south — for the first time.

Emperor Gia Long consolidated and developed Vietnamese nationhood during his reign from 1802 to 1819, but he and his two successors, Minh Mang (1820–1841) and Thieau Tri (1841–1847), were hardly visionary or progressive leaders. Supported by

OPPOSITE: Ceramic bas-relief at Wat Phnom, Phnom Penh, records the ceding of Battambang province to Cambodia in the 1907 Franco–Siamese treaty. ABOVE: French colonial architecture, Hanoi.

traditionalists, they ruled Vietnam with a whip in one hand and the rigid precepts of Chinese Confucianism in the other. They didn't like foreign missionaries, whom they regarded as a threat to the Confucian state, and when they began executing Vietnamese Catholics and expelling Jesuit scholars and priests they triggered a gathering clamor in France for intervention to stop the repression. It was when several foreign missionaries were put to death in 1858 that the French — who had helped put the Nguyen dynasty in power, after all — finally stepped in,

attacking Da Nang with a force of 14 ships and then moving south to capture Saigon.

As with most colonial expeditions, once the French toe was in the bath there was no stepping out. Spurred by Vietnamese resistance on the one hand, and efforts to open the society to trade and evangelism on the other, the French waded deeper and deeper into the region, first subduing the south then seizing the Red River Delta and Hanoi and finally imposing a Treaty of Protectorate on the imperial court in Hue. By 1887, the colonial takeover of Vietnam was complete — and, as subsequent events have shown, the seeds of the first Indochinese war were already sown.

It was not just Vietnam that had come under French rule; by this time Laos and Cambodia were also part of what the French proclaimed as the Indochinese Union. In Laos, the French had been welcomed as an ally against the Thais, and their legation in

Luang Prabang became the launching point of a campaign to push the Siamese right out of the country. By 1907, the Thais had pulled back across the Mekong River and Laos was a full French protectorate.

Cambodia's absorption into the union was inevitable, linked economically with Vietnam's Mekong Delta and providing a buffer, with Laos, against the kingdom of Siam. In 1863 the French persuaded the Cambodian monarch, King Norodom, to sign a treaty of protectorate, and 11 years later another agreement was coerced which gave the French full colonial power.

As harsh and unpalatable as it may now seem in this post-colonial age, French rule brought some benefits to Indochina. Colonial administration, though paying lip service to a series of puppet monarchs in each country, brought modernized government to the region and an infrastructure of roads, railways, communications, and institutions which is still quite clearly evident today. The French also put a stop to expansionism in the region, and this gave Cambodia and Laos their first real sense of security for centuries. While French military power kept the Thais at bay to the west, Vietnam's imperialist ambitions were also thwarted.

THE SEEDS OF REVOLT

It's probably because of this protection that both Laos and Cambodia gave the French very little trouble during what could be called the idyllic years of colonial rule. Resistance was minimal, and occasional rebellions were directed mainly at corrupt or harsh native officials rather than the French. But both countries paid the price of foreign occupation — their cultures and traditional institutions submerged by the weight of French administration, education, religion, and undeniable sense of superiority. While newly-emerging middle classes welcomed French ways, an underlying vein of anger and resistance obviously existed — ultimately exploding with brutal fury in April 1975 in the Khmer Rouge pogrom, directed at virtually everything foreign, sophisticated, or even educated, in Cambodia.

Vietnam was a much different kettle of fish. From the very beginning of French rule

ABOVE: Old colonial administrative building, Saigon. OPPOSITE: Saigon's most enduring landmark, Notre Dame Cathedral.

there was fierce and often widespread resistance — not surprising when set against the tremendous battles that had been fought through the centuries to keep China at bay. As in the other Indochina states, there were Vietnamese who came to terms with and even flourished in the transplanted French culture — particularly the ruling elite, bureaucrats, merchants and sections of the armed forces and police — but much of the vastly agrarian population found itself laboring under the yoke of a system set up to economically exploit the country and its people.

free to study in France and other Western nations eventually produced the one political force with the manifesto, dedication and organization capable of leading the nationalist cause: the Communists, under Ho Chi Minh.

After organizing strikes, unrest and uprisings against the French for more than two decades — and coming under increasingly fierce French reaction — the Vietnamese Communists found the opportunity they'd been waiting for when the Japanese occupied Vietnam and Indochina in World War II.

In times of rebellion, and there were many intrigues and uprisings, French reprisal was often brutal. The guillotine was imported to deal with extremists — one point on which the Vietnamese have never forgiven the French. Examples of these awful contraptions are on show today in the grounds of the War Museum in Saigon and in the Hoa Lo Prison Museum in Hanoi.

From the turn of the century right up until World War II, the French found themselves in almost continual struggle against nationalist movements. The Japanese victory over Russia in 1905, and the 1911 to 1912 republican revolution in China, heartened the dissidents. The rise of an urban proletariat and a growing corps of students who had been

Working with the Viet Minh — the name of these nationalist guerrillas taken from the League for the Independence of Vietnam (*Viet Nam Doc Lap Dong Minh Hoi*) — they were the only force to seriously resist Japanese rule. In 1944, the Viet Minh were receiving financing and military aid from the Americans, and thereby hangs a tragic tale: Ho Chi Minh thought he would get United States blessing for full independence after the war, and the Americans neglected an opportunity which would have saved both nations from the savagery of the Vietnam War in later years. In September 1945, with the Japanese defeated, the Viet Minh controlled so much of Vietnam that Ho Chi Minh was able to declare independence and proclaim the establishment of the Democratic Republic of Vietnam. What happened then was an act of almost obscene political cynicism on the part of the victorious Western powers.

The Chinese were given the task of disarming the Japanese north of the 17th parallel in Vietnam, and the British were assigned to the south. The Chinese went on a rampage in the north, and Ho Chi Minh had to accept French help, of all things, to halt their pillaging. In the south, the British not only used defeated Japanese forces to help maintain public order — it was soon clear that their main task was to help the French regain colonial power. The Americans stood by and did nothing.

INDEPENDENCE AND THE FIRST INDOCHINESE WAR

Within weeks, the French were ruling Vietnam again, but by now the lust for independence was too strong. In early 1947, some weeks after a vicious French attack on opposition elements in Hai Phong, serious fighting broke out in Hanoi and the first Indochinese war began.

It took the Viet Minh eight years to drive the French out of Vietnam for good, and the character and strategy of the combat was virtually a preview of the second, far more destructive, war to come. The French received massive American military aid. The Viet Minh fought a war of attrition, a guerrilla campaign, which combined shrewd military and political action — their political agenda based on growing disenchantment and opposition to the war in France itself. When victory finally came at Dien Bien Phu in May 1954 — the Viet Minh pouring in their thousands from trenches and emplacements after laying siege to the French positions for nearly two months — it was an ignominious end to the French empire in Indochina: the 10,000 or more soldiers and Legionnaires who surrendered were not just demoralized, but starving and virtually abandoned by their compatriots back home.

It seems astonishing today that the French fought so fiercely to retain Vietnam when they'd already divested themselves of the rest of Indochina by the time they made their last stand at Dien Bien Phu. Laos had been granted full independence in 1953 against the background of complex political maneuvering that threw up two nationalist figures who were to have a significant impact on the country's future course — Prince Souvanna Phouma, a neutralist leader, and the leftist Prince Souvanna Vong, who sought support from the Vietnamese Communists for a revolution under the banner of the Laos Patriotic Front, or Pathet Lao.

In Cambodia, another now-familiar figure had strode onto the national stage — Prince Norodom Sihanouk, proclaimed king by the Vichy French regime during the Japanese occupation and still enthroned when the country gained full independence along with Laos in 1953. Sihanouk later abdicated, allowing his father to become king, but virtually ruled Cambodia as prime minister and head of the all-powerful People's Socialist Community party until the monarch died in 1960. Then he made himself chief-of-state. Although a self-proclaimed neutralist, Sihanouk's conceited, self-centered policies — in which he treated Cambodia and its people as his personal property — created the seedbed, as we shall see, for left-wing revolution in Cambodia and the dark agony it ultimately suffered under the Khmer Rouge.

The full independence of Vietnam, Laos and Cambodia was officially ratified by the Geneva Accords of July 1954. But while these agreements left Laos and Cambodia intact, they divided Vietnam at the 17th parallel, with the despotic Ho Chi Minh's Communists ruling the north and an equally despotic, United States-backed anti-Communist, Ngo Dinh Diem, controlling the south. One of Diem's first official acts was to renege on the Geneva Accords, refusing to take part in nationwide elections which had been dictated in the agreements. He held a referendum instead, asking voters whether he should continue to rule the south. Amid claims of rabid vote-rigging, he won a resounding "yes" and proclaimed himself President of the Republic of Vietnam. The United States and most of its traditional European and Asian allies immediately recognized his regime, underscoring the

OPPOSITE: Young soldiers at Liberation Day parade, Da Nang.

17th parallel as the line that had been drawn against further Communist expansion in the region.

THE SECOND INDOCHINESE WAR AND THE AMERICANS

With Diem's rise to power, the die was cast for American involvement in Vietnam and inevitable war. Diem soon proved so corrupt, repressive and politically paranoid that he became an embarrassment to Washington and was overthrown and murdered in a

bombing reprisal against the North. And the Vietnam War — the second Indochinese War — had begun.

The Vietnam War lasted 11 years, and the terrible cost to both sides — the Americans and the Vietnamese — has been evident since the last shot was fired, and will be debated for decades to come. Its savagery deeply undermined the American public's faith in its own ideals and institutions and spawned a high-level political philosophy, espoused by a succession of presidents — Johnson, Nixon and, later, Ronald Reagan and George

United States-backed coup in 1963. But by now, Hanoi had launched its military campaign to liberate the south and reunite the country; Washington's bid to halt the toppling dominoes of Communist expansion had become a crusade, with Vietnam its chief battleground; and more than 16,000 United States military advisers, sent in by the Kennedy administration, were already trying to prop up a southern political infrastructure that was to remain every bit as repressive and corrupt as the Diem regime.

A year later, Kennedy was dead, two United States warships had come under an allegedly "unprovoked" attack in the Gulf of Tonkin off North Vietnam, and President Lyndon Johnson had launched a massive

Bush — that anything goes if it means containing communism. Defeat cost America its belief in its own righteousness and invincibility, much of its global prestige, and its supremacy in Asia, not to mention 58,000 servicemen killed in action and almost double that number who have committed suicide for various war-related aftereffects at home since.

For Vietnam, victory — as heroic as it was — cost an estimated four million civilian lives, along with hundreds of thousands of military casualties in the North and South. And while it reestablished national sovereignty and pride, it put the Communists in power, with the political and social repression and "cleansing" inevitable under hard-line, embittered revolutionary rule. More than that, it closed Indochina's doors to the rest of the world for some 15 years, during which a series of boilerplate Marxist social and economic experiments virtually bankrupted the three economies. In Vietnam's case, the American trading and financial embargo, imposed after the fall of the South in 1975, denied it access to desperately needed World Bank, IMF (International Monetary Fund) and ADB (Asian Development Bank) loans for reconstruction and development. If there is an epitaph to be written on the Vietnam War, it is the war that no one truly won.

We may well look back today with terrible fascination at the litany of events, misconceptions and mistakes that made the second Indochinese War such a global trauma. The American bombing campaign against North Vietnam, which started in 1963 and continued throughout the war, pitted high technology against a well-organized agricultural society, with relatively little industrial capacity, which was able to repair roads, bridges, dams and dikes as fast as they were damaged. When the United States Marines were first deployed in Da Nang in March 1965 — spearheading President Johnson's massive buildup of American forces in South Vietnam — the stage was already set for the ultimate failure of mechanized, main-force warfare against an elusive, highly mobile guerrilla enemy.

The United States and ARVN (Army of the Republic of Vietnam) rural pacification program, aimed at isolating the Communist Viet Cong guerrillas and destroying their infrastructure, simply led to the wholesale uprooting of thousands of rural families from their ancestral homes and left them to languish in armed resettlement camps. The strategy of search-and-destroy, an attempt to hunt down and eliminate Viet Cong and infiltrating North Vietnamese units, generally floundered in confusion — for who was to say with certainty who were Viet Cong and who were government loyalists in a population that all looked the same? Moreover, a CIA-operated covert war called the Phoenix Program which employed intimidation, terror, torture and assassination in a bid to fight the Communists on their own terms — and which virtually governed the pacification program throughout the conflict — got so badly corrupted and out of hand that it victimized and alienated millions of neutral Vietnamese.

The Americans lost the political war, too. Much of the failure of the Phoenix Program — as infamous as it was — has been attributed to the discreditable caliber of leadership that Washington supported, namely presidents Nguyen Cao Ky and Nguyen Van Thieu who are alleged to have used the pacification scheme more against their rivals than the Communists. The biggest political blunder

the American military hierarchy made was continually assuring an uneasy American public that the war was practically won, that the "light at the end of the tunnel" could be seen, as early as 1967. The Communist Tet Offensive of January 1968, however costly it proved in Communist lives, blew open the credibility gap between the Pentagon and the public, cut the heart out of American support for the war and turned the United States antiwar movement into a national cause. From that point on, American involvement in Vietnam was a prolonged tactical retreat.

OPPOSITE: Vietnam War flashback — panic at the United States Embassy in Saigon as Communist troops close in. ABOVE: Government tanks in Da Nang.

Perhaps the most disastrous step taken by the Americans was President Richard Nixon's invasion of eastern Cambodia in April 1970, an offensive aimed at destroying Viet Cong border sanctuaries and easing pressure on the United States in the Paris peace negotiations. Up until that time, Cambodia and Laos had been on the sidelines of the main conflict in Vietnam — though Laos had sustained an American bombing campaign estimated among the heaviest in history.

At the start of the American buildup in Vietnam in 1963, Laos had been in turmoil, with three political factions, including the leftist Pathet Lao under Prince Souvanna Vong and Prince Souvanna Phouma's neutralists, jockeying for power. In 1964, after a series of failed attempts to form coalition governments, Souvanna Vong pulled out of what he regarded as a rigged match and took his guerrilla forces into the mountains. American military advisers moved in to support the government, a massive bombing operation was unleashed against Pathet Lao bases and infiltration routes and the CIA raised an army of specially-trained Hmong hill tribe warriors and Thais for covert operations.

THE KHMER ROUGE HORROR

In Cambodia, United States involvement came about partly as a result of Prince Sihanouk's struggle to stay in power. In the early days of the war he declared Cambodia neutral, but then turned around and severed diplomatic relations with Washington and allowed North Vietnam and the Viet Cong to use Cambodian border areas as a sanctuary and infiltration zone. In 1967, he did another about-turn: facing a rural rebellion against his autocratic rule, and convinced that the Communists were out to get him, he began cracking down harshly on leftists. Two years later, the United States launched devastating B-52 bombing raids against guerrilla base camps in eastern Cambodia. In March 1970, Sihanouk was mysteriously deposed while on a trip to France by a rival faction led by General Lon Nol. Then he changed his political colors once again, taking up exile in Communist Beijing at the head

of a rebel Cambodian movement which he had dubbed the "Khmer Rouge."

Richard Nixon's invasion of Cambodia in April 1970 triggered savage warfare between the Khmer Rouge and Lon Nol's government forces. It raged for five years, and despite American military and economic assistance, the Lon Nol regime was no match for the rebels, led by a French-educated revolutionary whose name has since become synonymous with genocide, Pol Pot. When Phnom Penh fell to the Khmer Rouge on April 17, 1975 — two weeks before the Communist victory in Vietnam — these fanatical Maoist revolutionaries, characterized by their black peasant uniforms and traditional red checkered scarves, wreaked a barbarous revenge on their people.

One only has to visit Cambodia today to see the damage and horror that the Red Khmers inflicted during their demented four-year purge of Cambodian society. Proclaiming "Year Zero" as the start of a complete restructuring of Cambodian society, the revolutionaries virtually closed down Phnom Penh, forcibly evacuating most of its people to the countryside to work as slave labor on farming and rural development projects. Families were broken up, the parents separated and forced into work units that were often many kilometers from each other, the younger children placed in political education camps.

Meanwhile, thousands upon thousands of people were consigned to what has now become a morbid catchphrase of Cambodia's darkest hour — the "killing fields." Almost the entire intelligentsia and middle class were wiped out — dancers, writers, teachers, artists, anyone wearing glasses (they were taken as a sign of education). The victims, some of them mere children, were imprisoned, interrogated and brutally tortured before being taken to rural mass-execution spots where they were put to death — their skulls smashed with hammers and pickaxes to conserve bullets. There is no need to relate more about this wholesale murder, except to add that between one and three million people are believed to have perished at the hands of the Pol Pot regime. It takes just one visit to the terrible Tuol Sleng Genocide Museum in Phnom Penh — formerly one of the key

interrogation centers — and the genocide monument erected over one of the killing fields 15 km (nine miles) from the city, to sense the terrible chilling darkness that swept across the country at that time.

VIETNAM INVADES CAMBODIA

In December 1978, after a series of Khmer Rouge border provocations, Vietnam invaded Cambodia, mercifully putting an end to the pogrom. While it's highly unlikely that humanitarian aims triggered the invasion — the Khmer Rouge had been liquidating thousands of ethnic Vietnamese living in Cambodia — the outcome was another quite shameful show of international cynicism. China and the United States, both alarmed at the prospect of Vietnamese expansion in Indochina, supported the Cambodian rebel factions, including the Khmer Rouge. Thailand, equally nervous about Vietnamese power, became the sanctuary for Red Khmers and other rebel groups forced westward by the occupation.

The Vietnamese behaved relatively well during the 11 years they occupied Cambodian regions, providing a period of security in which the country could recover from the trauma of the Khmer Rouge "revolution." In 1989, under tremendous pressure to mend its political fences with the rest of the world — particularly the United States — and reopen its society to foreign aid and investment,

A candlelit procession around Vientiane's most sacred stupa climaxes the That Luang rites.

Vietnam withdrew its forces from Cambodia. The United States immediately severed its support for the Khmer Rouge and instituted its so-called "road map" of moves — orchestrated in concert with the hunt for American MIAs, or servicemen still missing in Vietnam — to establish diplomatic relations with Hanoi.

THE AFTERMATH

In September 1990 the United Nations Security Council instituted what was hailed as a major step toward eventual peace and stability in this war-weary nation — setting up a Supreme National Council in which a coalition of Vietnam-backed government figures and the major rebel groups, including the Khmer Rouge, would run the country while preparations were made for national democratic elections. The interminable Prince Sihanouk returned in triumph from Beijing to head the council, and immediately starting plastering huge portraits of himself all over Phnom Penh. Some 15,000 United Nations troops and rear echelon from nearly 30 countries were deployed in Cambodia to help rebuild the country's infrastructure,

ABOVE and OPPOSITE: Four faces of Vietnam — "beginning to enjoy life again."

disarm the rebel factions, resettle thousands of refugees from Thailand and supervise the May 1993 elections.

The elections in mid-1993 saw Sihanouk seeing to his own interests as usual: he was appointed king and his son prime minister. The elections were predictably boycotted by the Khmer Rouge, consensus not being rel-

evant in their mode of government. Allowed to pour back into Cambodia, they dug in again, controlling virtually all the northern half of the country. Backed by Thai political and business interests deeply involved in lucrative logging and gem mining in their territory, they launched increasingly daring attacks on United Nations positions. More than that, they continued the pogrom that they had launched in 1975, executing and assassinating whole communities of ethnic Vietnamese in their area.

Yet with the death of Pol Pot in 1998 and the near demise of the Khmer Rouge, it could be time for a new era to open up for Cambodia. However, after the 1998 elections the country plunged even deeper into chaos, as Prince Norodom contested the winner Hun Sen, saying the elections were rigged. Riots and demonstrations broke out around Phnom Penh with attendant confusion. Perhaps it is this country's fate to remain in turmoil, with or without the added insult of the Khmer Rouge.

INDOCHINA'S PEOPLE

At the crossroads between Southeast Asia and China, Indochina has a scintillating mixed heritage, with numerous minority ethnic groups in addition to each of the country's main inhabitants. Millenniums of fighting, migrations, trade, and different rulers presiding over waxing and waning empires have resulted in an enormous variety of faces and cultures.

Vietnam alone has within its estimated population of 73 million, 54 ethnic minority groups. While the Viets comprise something around 85% of the population, the ethnic minorities fall into three main groups. The

northern minorities, or *Montagnards* as they were called by the French, are of the Austro-Asian family, who migrated from the tribal lands of southern China and encompass the Hmong, the Zao, the Tai-Kadai, the Tai, the Nung and the Muong. In central Vietnam are the Austronesian groups which include the Jarai, the Ede, the Ria Gia and the Cham, who live around Phan Thiet and also on the Cambodian border near Chau Doc. Thirdly, smaller communities of Sino-Tibetan heritage, encompassing such obscure groups as the La Hu, the Hoa, the Ha Nhi, the Lo Lo, the Si La and the Phuia, live in the far north-western tip of the country where it borders Laos and China.

Laos, with a population of around 4.7 million, comprises 47 ethnic groups, many living traditional lifestyles high in the mountain areas bordering China and to the east, bordering Vietnam. In the south a large percentage of the population is made up of the Laos Loum (lowland Laos), Laos's main population group.

Existing alongside the different groups are the Chinese, estimated at around five percent, who have come over the centuries for trade, and more recently as imported labor from southern China, especially to the northern provinces such as Oudomxai.

Cambodia has a population of around 10 million, comprised of 90 to 95% Khmers — whose likeness can be seen in the huge, enigmatic carved stone faces of the Bayon Temple at Angkor Thom. The Khmer race is related to the Mons of Burma and Thailand, migrating, it is believed, to what is now Cambodia over 4,000 years ago, and inter-marrying with the existing Austronesian population. The remaining five to ten percent include Chinese Khmers, ethnic Vietnamese

who live on Cambodia's two main rivers and the Tonle Sap Lake, and the approximately 500,000 Chams or Islamic Khmers, descendents of Vietnam's great Champa Kingdom, who live scattered across the southeast. In the northeastern highlands close to the Lao and Vietnamese Borders are animist tribal groups known as Khmer Loeu who share their ancestors with those across the border.

THE ECONOMY

With Vietnam at its fore, Indochina is standing on the threshold of an economic boom. The region is rich in agricultural, marine, forestry and mineral resources, and war and austerity have kept all these treasures intact. It has a huge pool of educated but cheap labor ready to offer highly competitive offshore manufacturing for the established export economies of Asia and elsewhere.

Both Laos and Cambodia have followed Vietnam's lead and opened up their economies, with Laos launching the same sort of campaign for foreign investment in 1987. But where Vietnam was shackled by the United States trade and investment embargo, Laos was given Washington's blessing the moment it opened its doors and has been receiving Asian Development Bank (ADB) loans for several years. In fact, foreign aid is estimated to make up more than 75% of the national budget — not surprising when you see that Vientiane is awash with nongovernmental organizations of all descriptions — as one observer noted, if all the money spent on aid workers and their big-budget lifestyles was spent on Laos, there would be enough wealth to last the country twenty years.

Indochina is hungry for consumer goods, and for investment funds to build an industrial base and redevelop its neglected infrastructure. It sees itself as a future Pearl River Delta, copying the phenomenal manufacturing boom that's taken place in China's southern special economic zone.

If there's any question about the region's future, it's just who is going to lead this dramatic economic renaissance. In this respect, history is repeating itself: Vietnam, flexing potentially powerful economic and political muscle, sees itself as the dominant partner in a new Indochinese economic zone; but to

the west, the Thais are again claiming cultural and economic suzerainty over Laos and Cambodia. And the outcome of this struggle may well hinge on what happens in Cambodia in the coming years.

In the meantime, Vietnam's plans for economic development can be taken as a blueprint for the entire region. Offering some of the most attractive investment regulations of any developing socialist nation — including 100% foreign ownership of selected enterprises — it has charted a course which it hopes will bring it to economic "takeoff" by

the year 2000. By then, it would be sustaining an annual growth of seven to eight percent and a per capita income of around US$500 and be ready to graduate to NIE (Newly Industrialized Economy) status.

Hanoi's economic architects, led by experienced southerners, have taken tourism development as the first step, bringing in much-needed foreign exchange. Then the "transformation industries" — foreign joint-venture manufacturing based on textiles, electronics and other export products — would take over, creating an industrial base. Agricultural production would be boosted

OPPOSITE and ABOVE: The motorbike and bicycle remain Vietnam's main form of personal transportation as the country tries to modernize.

and modernized, and it's this sector which is most vividly reflecting the country's enormous economic potential. With the introduction of free enterprise and the removal of barriers to the flow of produce between provinces and cities, Vietnam has in a few short years already progressed from a food-deficient nation to the world's third largest exporter of rice.

Much of its hopes are pinned on oil production from a series of former Soviet-operated offshore leases in the Gulf of Tonkin and the South China Sea off central and south-

Vietnam has most definitely been viewed as the new frontier of business and investment in Asia. As one enthusiastic Western oil executive in Saigon remarked: "We see tremendous potential here, and probably better prospects than those in other countries of the region, including Thailand." But by 1998, another observer remarked that the initial enthusiasm has dimmed, and one in two foreign investment projects are being dropped in Hanoi, and one in four in Saigon — a disappointing average.

ern Vietnam. Now transferred into some of the world's leading multinationals, including BP Petroleum, these fields produced four million barrels of crude in 1991 and the plan is to top 10 million barrels a year by the end of this decade. No major new strikes had been made at the time of writing, but the feeling among foreign executives was that the southern fields, in particular, could prove to be a new North Sea.

The biggest drawback to Vietnam's development continues to be its decrepit infrastructure — roads, bridges, ports, airports, transportation and communications — a problem that is only now being resolved as one of the results of the lifting of the United States embargo.

The Thais are one of the region's most enthusiastic investors, digging in early in both Cambodia and Laos, although one particular industry has embroiled Bangkok in a potentially bitter political and environmental controversy — logging.

Both Laos and Cambodia are primarily agricultural economies, though they also possess extensive mineral deposits which require heavy investment to exploit. While the Lao Government has instigated some strict and sustainable anti-logging laws, they are not always implemented. Amongst poor, underpaid officials, a juicy bribe is sometimes

ABOVE: New fishing trawlers are being built as the economy revives. OPPOSITE: Cambodian student, Phnom Penh.

difficult to refuse. But that said, a former governor of Attapeu Province was sentenced to fifteen years jail for his part in a logging scandal.

The Thais, having already created an environmental crisis with their own wholesale deforestation, have virtually seized upon this opportunity as their own, and Laos is particularly vulnerable—only 10% of its land is arable, the rest is forest. The construction of the Friendship Bridge across the Mekong at Nong Khai, donated by Australia and opened in 1993, is seen by some environmen-

talist and economists as not just the country's first real physical link with the outside world, but an opportunity for increased Thai exploitation as well. When a second Japanese bridge linking Pakse in the south with Thailand is completed, far away from the stern eye of the central government, opportunities for illegal logging will be vastly increased, no matter what good intentions the government tries to implement.

In Cambodia, Thai logging operations took a more cynical and sinister twist — creating a business partnership with the Khmer Rouge. It's from the Khmer Rouge territories, close to the Thai border, that most of the teak, along with other valuable timbers and gems, were being extracted. It's not just causing environ-

mental concern, but the possible unchecked rape of Cambodia's remaining forests.

POLITICS

In spite of all the new political openness Vietnam is still a socialist country, its people still forced to toe the Communist Party line, its economy still centrally controlled. It has opened its doors, but at the higher echelons of government it remains suspicious of foreign activities and motives. Add to this the negatives of greed and corruption and the picture continues to look bleak. But while political change is dragging its feet in the wake of economic liberalization change is certainly taking place.

The *Doi Moi* open door policy of 1986 came about because a liberal southerner, Nguyen Van Linh, assumed power at the Sixth Congress of the Communist Party. Once the door was opened, once the fundamental decision to bring back free enterprise was made, it would have taken a reactionary crackdown similar to China's Tiananmen Square massacre—especially in the south—to turn back the clock of change.

The switch to a market economy has meant that a whole gamut of laws and regulations have had to be rewritten, and new ones made, particularly with regard to that anathema of the Communist state, property ownership. The influx of foreign investment has meant the liberalization of banking laws, allowing private and joint-venture banks to compete with the previous monopoly of official institutions. The search for export markets has meant that Vietnamese are now allowed to travel overseas. Thousands of detainees were released from reeducation camps, and some prominent intellectuals and former capitalists rehabilitated, as the state casts about for people capable of guiding and managing economic reconstruction.

Moreover, the state itself has had to change. Immediate decision-making power is said to have switched from the unwieldy Communist Party politburo to a smaller and more decisive executive council, with each member responsible for a particular arm of government and administration.

Another trend has been presented by one of the leading figures involved in the reform program, a prominent Saigon business consultant and former Deputy Prime Minister of the wartime South Vietnam. "What we're aiming for is market socialism," he says, "a free-market economy and society with firm, but not dictatorial, guidance and leadership from the top. We really don't think that Western-style moral democracy suits Vietnam at the present time. We've seen how South Korea and Singapore achieved spectacular economic growth, both of them under firm leadership, and we feel this is what Vietnam needs to accomplish its prime task — to catch up and compete effectively with the rest of the world."

There are two other clear trends that point to comparatively radical political change in the future. The demand for talented, foreign-trained technocrats to run the economic program means that power will ultimately shift from the surviving old guard of wartime revolutionaries. And it remains to be seen how long the Vietnamese will be content with the power to conduct their own business affairs but otherwise not think for themselves.

In Laos, the death of President Kaysone Phomvihane in November 1992 seemed to herald the same passing of the Communist old guard in favor of younger, more liberal blood. However, his successor, Nouhak Phoumsavan, is a similarly hard-line Communist and ally of Vietnam, underscoring the careful snail's pace of change in Laos and Hanoi's struggle to maintain its political grip on the country in the face of economic incursions by the Thais.

But while Phoumsavan's accession maintains the supreme power of Laos's People's Revolutionary Party, or Communist Party, governing through its politburo and central committee, an undercurrent of economic reforms similar to those in Vietnam has served notice of the course the country will no doubt take in the future. The first legal code since the revolution was enacted in 1988, establishing a Western-style system of courts and justice, along with liberal investment laws. In 1990, Laos was given its first official constitution since the Pathet Lao victory. It not only ratifies the principle of free enterprise but also removes the term socialism and the hammer and sickle from the nation's political banner.

The Laos government is regarded as one of the most secretive in the world. Access to its 12 ministries is not easy for outsiders, and media publicity is rare. But it is known that Laos shares Vietnam's prime problem — finding the experience and talent needed to build and manage a modern economy. Otherwise, there's hardly any overt evidence of repression or interference in everyday Lao life; in fact, the society appears to be the most serene and idyllic in Asia. There seems to be complete religious freedom in this devoutly Buddhist domain, and when you witness the

passion with which the Laos celebrate their Buddhist festivals the term "Buddhist socialism" springs to mind. You can't help speculating that this is one popular power that the Lao Communists, at least, prudently came to terms with, instead of trying to conquer.

The same cannot be said of the Khmer Rouge in Cambodia. During their four-year rampage they destroyed Buddhist temples, murdered monks, and virtually eliminated all social and political institutions in their crazed vision of a new agrarian society which, if it had been Europe, would have turned the clock back to the darkest of medieval times. Much of Cambodian society has had to be completely reconstructed — and much is still in waiting. The country as yet has no real legal system, or even workable laws in place.

Where Vietnam and Laos are resisting democratic reform, it has been the linchpin of United Nations efforts to rebuild Cambodia's society. It's not just the political

OPPOSITE: Cyclo driver and typical load, Da Nang.
ABOVE: Poster in Hue recalls Vietnam's postwar revolutionary era.

framework that's had to be restored — just about everything from public security to social services has required building, virtually from the ground up. But it's in the political sphere that the most dramatic reforms have taken place. Since the mid-1998 elections and their crucial significance for Cambodia's redevelopment, more than a dozen political parties representing liberals, conservatives, pressure groups, and Buddhists have been jostling for a place in the new political order. Of the three main rebel factions, the Khmer Rouge did not field can-

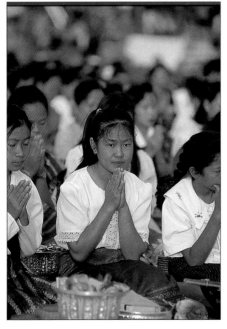

didates. A kingdom has been established under Sihanouk, nominally a constitutional monarchy, but in which he may well forget the word "constitutional," unless some remarkable change has overtaken him; his two sons have squabbled over the extent of their own influence, and one was prime minister until ousted and banished from the country by the current prime minister, Hun Sen. But Sihanouk, always a catalyst in any political shift in his long-suffering country, is now an old man, after 40 years of political skullduggery, and is also unwell. Cambodia's future is far from certain, the controversy over the results of the guaranteed "free and fair" 1998 elections leaving us to wonder about the future.

RELIGION

Multitudes of cultural antecedents and migrations have formed a dazzling mosaic of cultural patterns and beliefs in Indochina. Basic animistic beliefs, common to each country, formed the basis upon which were overlaid a veneer of Hindu and then strong Buddhist beliefs, which in Vietnam have given way to the mass faith of communism.

When China launched its own open door policy after Mao's death, following more than 20 years of irresponsible social and economic engineering, it got to work immediately rebuilding and renovating damaged temples, a trend that stemmed not so much from a restoration of religious freedom as from the need to give tourists something to see. Then the monks and worshippers had to be allowed back to give an air of authenticity. Such cynicism does not necessarily apply to all religious restoration, but it is undeniable that tourism development and the demand for exoticism is playing a major role in the revival of traditional culture right through the socialist world.

In Laos and Cambodia, the revival has been more popular, and more spirited, reflecting the fact that these communities have been under Communist discipline for a comparatively shorter period. While it's true to say that for every Buddhist or Catholic priest there is a party cadre keeping tabs on the congregation, there's been a wholesale rush back to the images and altars that puts Indochina's socialist experiment into proper perspective — a slight deviation along a well-trodden cultural path that stretches back many centuries.

We've read how Hinduism was a major spiritual force in Indochina, with the region's two greatest kingdoms, Champa and Angkor, profoundly affected by the cult of the *devaraja*, or god-king, in which the ruler acquired divine status. The Viets brought Confucianism from China — notwithstanding Confucius' mutterings about them — and this, together with Taoism, percolated down from the north. Confucianism is more a sys-

ABOVE and OPPOSITE: Buddhism has survived the Marxist revolution in Laos, proving to be a powerful cohesive force.

tem for the ordering of social responsibilities and relationships than a religion, while Taoism concerns itself with the individual's harmony with nature.

In the second century, a new spiritual influence, Buddhism, began sweeping the region, making its way from India through China to the Red River Delta and then down through the kingdom of the Chams. It eventually became the dominant faith in Indochina, adopted as the state religion by the Ly Dynasty in north Vietnam in the years 1010 to 1225, and spreading from there.

Although it came under Confucian counterattack in later years, Buddhism prevailed by coming to terms with other faiths around it — most notably, Taoism and Hinduism. One of the region's most precious religious sites, the ruins of Wat Phu near Pakse in southern Laos, is regarded as the first Buddhist temple in Southeast Asia. Yet its architecture is distinctly Khmer Hindu, reflecting the marriage that Buddhism was able to secure with the monuments of original faiths. The great Hindu city and temples of Angkor had likewise switched their religious focus from mainstream Hinduism to a Buddhist culture by the thirteenth century.

The arrival of the French with their Jesuit missionaries brought Roman Catholicism to the region, and this too eventually flourished — strengthened and promoted by colonial rule, especially among the more cosmopolitan, urban Indochinese. If you look at the general religious profile of Indochina today there are surviving outcrops of Christianity — some of the most beautiful old churches and cathedrals you can imagine — in a great sea of Buddhism, with Taoist, Hindu and Muslim temples here and there, most erected by immigrant traders who flocked to the region during the reign of the French.

While Catholicism and Buddhism have managed to coexist quite harmoniously, their relationship hasn't been without its upheavals. When Ho Chi Minh's Communists gained power in North Vietnam in 1954, nearly one million Vietnamese Catholics fled to the south. There they established a Catholic, anti-Communist oligarchy headed by Ngo Dinh Diem and successive presidents and supported by the Roman Catholic Church in the United States. Their power,

not to mention the corruption and favoritism that marked the wartime regimes, provoked and alienated the Buddhist majority in the south, triggered strikes and student unrest and led to the horrifying, internationally publicized self-immolations by Buddhist monks that led to Diem's downfall.

Not surprisingly, Catholicism came firmly under the official thumb in Vietnam after 1975. Many churches were closed and congregations intimidated. But a general restoration of religious freedom has taken place since the moves to liberalize the society began.

Right through Vietnam, old churches and cathedrals are now filled with worshipers each Mass — Saigon's Notre Dame; the striking, almost medieval Saint Joseph Cathedral in Hanoi; the beautiful, domed Notre Dame Cathedral in Hue, just to name a few. But when you take into account that only about eight percent of Vietnam's population is Christian, the revival of Buddhism and other beliefs is far more dramatic. Prayers, joss and ritual offerings are an everyday scene again at Taoist and Buddhist temples and shrines in all the major Vietnamese cities — places like the Tran Quoc and Quan Thanh temples around Hanoi's West Lake; the revered Thien Mu pagoda and temple overlooking the Perfume River in Hue; and two particularly evocative temples in Saigon, the tiny but elaborate Chinese Thien Hau Pagoda in Cholon, dedicated to the Goddess of the Sea, and the Le Van Duyet Temple three kilometers (just under two miles) from the downtown area.

But nowhere in Indochina is Buddhism celebrated with such devotion as in Laos, where the faith seems to have stood as a social bulwark against the excesses of Marxism. The great temples of Vientiane and Luang Prabang — and there are literally dozens of them — are flourishing again, apparently undamaged by the Communist reign; and, with new supplies of cash flowing in, some are being renovated. In fact, the only damage seems to have been committed in the war years before the revolution, when priceless images and relics were stolen by United States and other foreign correspondents, servicemen, and art dealers.

A sacred white elephant is paraded through Vientiane during the That Luang festival.

The extent to which Buddhism has triumphed in Laos can be witnessed at two major festivals each year. In November, thousands of monks and novices flock to Vientiane to join the population in a tumultuous three-day celebration centered on the country's most important Buddhist monument, the towering Pha That Luang (Great Sacred Stupa) and monastery on the city's northeastern fringe. In April, elephant processions highlight the three-day lunar new year festival in Luang Prabang.

In Cambodia, Buddhism and other faiths are emerging from a period of wanton destruction waged by the Khmer Rouge. Christian churches were completely destroyed during their rampage, and one of the country's most sacred Buddhist monuments, the hilltop temples and mosque of Udong, 40 km (25 miles) north of the capital, bears evidence of the suppression — the two main temples are in virtual ruins, along with huge reclining and sitting Buddha images which were dynamited by the revolutionaries. For all this, much of the country's Buddhist heritage is reasonably intact — most notably, the elaborate Royal Palace in Phnom Penh, a complex of halls and temples fashioned in the style of the Grand Palace in Bangkok, which the Khmer Rouge are said to have preserved in an effort to bolster their international image.

Of all the religions of Indochina, one bears special mention for its unique color and ritual — the Cao Dai of southern Vietnam. Based in Tay Ninh, northwest of Saigon on the Cambodian border, Cao Dai has a priesthood modeled on that of the Roman Catholic Church, and a doctrine which borrows much from Mahayana Buddhism, but otherwise combines beliefs from all the world's major religions. It was founded in 1926 by a Vietnamese civil servant, Ngo Minh Chieu, and most of its early followers were Vietnamese bureaucrats working in the French administration.

The Cao Dai virtually ruled Tay Ninh Province and parts of the Mekong Delta in their early days, and resisted the Communist Viet Cong during the Vietnam War. They came under revolutionary suppression after the war, but are flourishing again today — and you have only to visit their ornate Great Temple in Tay Ninh to realize why. With Masses celebrated four times a day, the vast congregations resplendent in white robes and the almost medieval costumes of priests, cardinals and other clergy, the Cao Dai have become one of south Vietnam's biggest tourist draw cards.

GEOGRAPHY

Indochina is a physically beautiful region, and one of many remarkable contrasts. A fleeting overview would compare the dra-

matic offshore karst, or limestone, formations of Ha Long Bay, east of Hanoi, with the vast, dazzling sweep of flooded rice-plains in the Mekong Delta; or the crumpled, green-swaddled folds and sharp jungle-covered peaks of the mountains of northern Laos with the tranquil surrealism of Cambodia's southern rice lands — flat green pastures broken with the distinctive outlines of tall sugar palms.

Each country has its own contrasts. Vietnam, stretching more than 1,600 km (990 miles) down the eastern seaboard of Indochina, is virtually two huge rice-bowls — the northern Red River and southern Mekong Deltas — connected by a spine of jungle-covered mountains, the Annamite

Cordillera, which form the Central Highlands. Within this general physical profile, other distinctions appear: long swathes of sandy untouched beaches which stretch one after the other from north to south; the pristine coastal bays and flat rice fields of the region between Da Nang and Hue, set against the rising foothills and high mountain passes of the Central Highlands; the largely arid central region of the country, beyond Quang Tri and what used to be the Demilitarized Zone (DMZ), compared with the lush riverine landscape and rich farming communi-

bleak, red-soiled terrain that's reminiscent of outback Australia minus the eucalyptus.

Cambodia is basically a broad and flat central alluvial plain, densely populated and fed by the Mekong River, with hills, mountain ranges and escarpments to its southwest, north and east. The broad expanse of the shallow Tonle Sap, which spreads below the Angkor region towards Phnom Penh, is a great provider of fish and marine life, reputed to have more fish per cubic meter than almost any other water body in the world. The Mekong Delta begins in this central region,

ties that lead into the delta area south of Saigon.

Laos is far more rugged, with nearly three-quarters of its landlocked terrain covered with mountains and plateaus, networked with clear water rivers that rush west to the Mekong. Some of the hills are more than 2,000 m (6,500 ft) high in Xieng Khuang Province, home of the Plain of Jars. In stark contrast, the Mekong River Valley around Vientiane and Savannakhet is a flat, fertile plain, and the source of most of the country's food. The mood of the landscape changes dramatically between Vientiane and the upland regions — sultry farm lands, dotted with palms on the one hand, changing in the Xieng Khuang area to a harsh, somewhat

and the river is several kilometers wide in some areas. It splits into two separate courses, the Mekong itself and the Bassac River, at Phnom Penh before sprawling eastward into southern Vietnam. Forests cover the Eastern Highlands and hills in the southwest, close to the Thai border, where the Khmer Rouge have been operating their lucrative lumber trade with the Thais. The southern coast, facing the Gulf of Thailand, is the country's future tourism drawing card, a potential resort playground of sandy palm-fringed beaches and islands that look like tropical atolls from the air.

ABOVE: Open-air cafe on the Mekong River in Vientiane — an excellent spot for viewing sunsets.

Northern
and
Central
Vietnam

NORTHERN VIETNAM HOLDS A SPECIAL fascination for visitors to Indochina — a mysterious and forbidden citadel throughout the Vietnam War and the years of isolation that followed. For decades it was a society that most people could visit only in the imagination, piecing together sketchy media images and accounts of revolutionary rallies, United States bombing statistics, peasant work units rebuilding bridges and dams, and human ants struggling in their thousands along the jungle of the supply lines of the Ho Chi Minh Trail. And Ho Chi Minh himself, his wispy, bearded features symbolizing the stubborn heroism of a society which is only now showing what it actually cost to win the war.

What it cost was progress. Aside from some uninspiring Soviet-era construction — mainly banks and other public buildings — development throughout the north has stood still for more than 40 years, leaving the society to muddle through with an antiquated infrastructure left behind by the French. The north is poorer and more rustic than the south, its industrial capacity comparatively ramshackle, its roads in chronic disrepair, its bridges barely able to handle today's traffic, let alone what's likely to come as the region modernizes.

On the one hand, this predominantly agricultural society illustrates the futility of the Vietnam War—how fundamentally reckless it was to wage a high-tech bombing campaign against rural roads, dikes, bridges, grain storage facilities and rail-lines that could be bandaged back together as fast as they were damaged. Even today, the railway bridges serve a dual purpose as one-way vehicular bridges between trains. Controlled by traffic lights they illustrate the brilliant Vietnamese capacity to maximize use of valuable properties. But what's more illuminating is the social character of the north — so reserved and disciplined, dignified and yet so hospitable, whether urban or rural — and what a tragically wide gulf existed between these people and the vulgar Texan bravado that characterized President Lyndon B. Johnson's ill-fated obsession with bringing them to heel.

And as I've mentioned before, there's a rustic, historical charm to the north that the war and isolation have left in their wake —

an architecture that's languished in a time-warp since 1945, a cultural naivete that's somewhat refreshing amid the increasingly hard-nosed growth-driven economies of East Asia.

There's already a well-worn tourist route between Hanoi, Hai Phong and beautiful Ha Long Bay, while access to the rest of northern Vietnam has opened up to reveal destinations of great beauty. It is true that some of the roads need work but for the wonders and magnificent scenery in the northern mountain ranges revealed, a little discomfort is a

small price to pay — after all, Indochina is Asia's last great adventure.

HANOI

Stately, sedate and oh so polite, Hanoi makes an interesting introduction to Vietnam. Dun-colored buildings line the streets, the drab shades only relieved by splashes of red from the ever-present flag with its bright yellow star and the scarlet hatbands of the omni-present guards. Anyone sensitive to atmosphere will notice a slightly oppressive air in Hanoi, a restrained kind of tension, a "big brother is watching" sensation, and perhaps

Hanoi's Tran Vo Temple OPPOSITE and Buddhist temple-keeper ABOVE.

even experience a mild desire to look over your shoulder. Quite often, local people speaking with foreigners *do* tend to look over their shoulder, perhaps a flashback to the old days. The atmosphere is much more noticeable if you are arriving from freewheeling Saigon, so far to the south.

One thing that astonishes any visitor to Hanoi is the extent to which the capital's French colonial heritage survived the Vietnam War and, even more remarkably, survived the onslaught of developers since the war — so much so that it seems there is hope

Densely populated Hanoi sprawls for miles in all directions, although most of the sights to visit are located in the relatively small city center. Streets are laid out in a simple grid system created by its French planners. The four principal lakes make convenient reference points for the various districts and visitors will find most tourist sights within the boundaries set by the lakes.

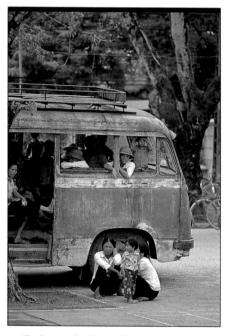

the historic parts of the city, and especially the 36-Streets district west of Hoan Kiem Lake, may remain. Although much of the architecture is neglected and crumbling, in the summer months particularly, when the city's tree-lined boulevards are in full leaf, Hanoi exudes an exotic, early-European character which you'll not find anywhere else in Vietnam.

Many of the old villas are being renovated for use by big business concerns, such as the ANZ Bank near the Hoan Kiem Lake, and upmarket residences that have been taken over and renovated for the executive managers of major companies. A series of inner-city lakes and small parks, and its location on the Red River, add to the rustic beauty.

To the north, Hoan Kiem Lake is the tourist mecca. Located nearby are most of the newer hotels, galleries, tourist shops, and most important of all, the magical Old Quarter and the main market area around Silk Street and Dong Xuan Market.

Continuing northwest you arrive at West Lake (Ho Tay), the site of the newer big hotel complexes and business developments. En route you pass by the finest surviving examples of the colonial culture — streets full of old French villas, and the imposing and monumental neo-socialist construction and vast square of the Ho Chi Minh Mausoleum.

To the south, Thien Quang and Bay Mau Lakes are the center of the city's "new" business district, where foreign and local companies are clustering their offices in renovated commercial blocks and villas.

From a tourist's point of view, the city's key thoroughfares are Hai Ba Trung Street, (close to Hoan Kiem Lake), running east-west, which has become a mecca for imported consumer electronics; Ngo Quyen Street, running north-south, where you'll find the Sofitel Metropole; and Trang Tien Street which leads east-west from the Opera and new Hilton Hotel down to numerous restaurants and galleries before veering north

(right) into Pho Dinh Tien Huang, site of the main Post Office, and Lake Hoan Kiem.

To the north and west of Hoan Kiem Lake is the Old Quarter. West Lake is further northwest of the main city center. To the city's northeast, the antique but elegant Long Bien Bridge, built by the French, is one of the city's main cross-river accesses to the international airport and the road to Hai Phong and Ha Long Bay; it's packed most of the day with thousands of rural people flocking in to sell produce and buy supplies in the markets. Remember that in Hanoi, major streets (as well as smaller ones) change names every few kilometers or so. This can make getting around extremely disorienting, and it is easy to lose your direction unless you are well-

armed with a decent map to help to keep your bearings.

Although all the major hotels have cars, minibuses, and limousines at their disposal, there is really only one way to get around Hanoi—by cyclo. These open contraptions, something akin to a front-end loader powered by a bicycle, are exceedingly comfortable, and are perfect for a city in which the people, the noise, the smells and the passing kaleidoscope of tree-fringed shop-houses, markets and villas lose their impact and exoticism from behind the darkened glass of an air-conditioned car or van. The cyclo drivers are primarily ex-war veterans, most of them educated and polite, and they'll take you anywhere you want to go. The cost runs about US$5 per hour, and hard bargaining is required. Select a driver you particularly like and hire him on an exclusive standby basis at US$20 a day. Some of the cyclos have seats wide enough to take two people, which makes cyclo-touring in Hanoi especially cozy for couples.

Motos, or motorbike taxis, are great for getting around and cost about US$1 a trip, although short trips can be bargained down to 50 cents if you want to try. Some hotels arrange for guests to rent bicycles, and you can also ask about renting a motorbike.

Taxis are clean, metered and new. One company I found particularly good is Hamatco ℭ 826 4444. The driver spoke English well enough and the service was faultless.

GENERAL INFORMATION

Hanoi was once dominated by the government-run Hanoi Tourism and Vietnam Tourism, but now there are plenty of private companies to choose from. One of the better run private companies is **Especen** ℭ (4) 826 6856 FAX (4) 826 9612, 79 Hang Trong Street, a stone's throw from the Vietnam Airlines office, while **Vietnam Tourism** ℭ (4) 826 4154 FAX (4) 825 7283 is at 30A Ly Thuong Street. The budget price **Green Bamboo** ℭ (4) 826 8752 FAX (4) 826 9179, 42 Nha Chung, is also a popular choice for short tours.

OPPOSITE: Hanoi market vendor LEFT and ramshackle public transport RIGHT. ABOVE: Cyclo drivers still provide city's main transportation.

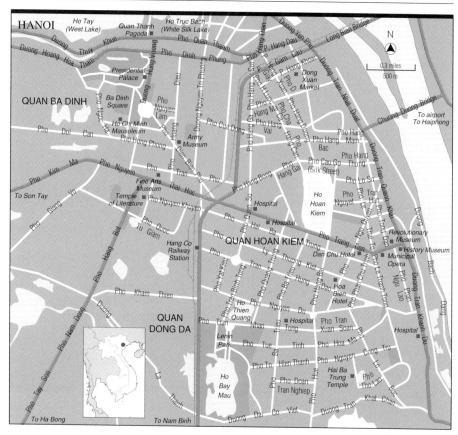

WHAT TO SEE AND DO
——————————————

To make it possible to keep some kind of bearings, I suggest starting with places of interest in the Hoan Kiem District where most sights are within easy walking distance or a short cyclo ride away from the Lake, while the others require a short taxi ride or, for the adventurous, a speedy, cheap and sometimes terrifying *moto* or motorbike taxi.

Hoan Kiem District

In Hoan Kiem District, the **Old Quarter**, a teeming rabbit warren of narrow streets and old colonial shop-houses, undeniably ranks as Hanoi's most interesting sight. Lying north of Hoan Kiem Lake, it is Hanoi's main market area and most populous district. Also known as the **36 Streets**, the district became the center of 36 guilds or *dinhs* centuries ago, when craftsmen came in from the surrounding countryside, each guild forming its own community in one area, hence the names of

their streets — Silk Street, Paper Street, Rice Street. What makes this very traditional area so compelling is that it is still composed of communities *in situ*, living a similar lifestyle as a century or two ago, in spite of the profusion of motorbikes, satellite antenna dishes and other signs of burgeoning wealth.

Within the 36 Streets is a mix of historic temples, guild and clan houses, colonial villas and the so called "tube houses," some of which are no wider than two and a half meters (eight feet) across, but tunnel back 100 m (300 ft) or more, with a rabbit warren of living quarters off the main tunnel or "tube." Walking around, with or without a map, reveals fascinating discoveries at every corner. You will find ancient craftsmen making wooden drums the way their forefathers did. Bamboo items line Bamboo Street. Paper Street (Hang Ma) is filled with shops selling

OPPOSITE: Ngoc Son temple bridge on Hanoi's Hoan Kiem Lake TOP, and BOTTOM the entrance to Quan Thanh Pagoda.

the handmade paper favored by artists and artisans, which also makes exclusive wrappings for gifts. Silk Street (Pho Cau Go) once housed silk merchants, although recently it has lost much of its authenticity, if not its charm to tourism. Today it is filled with galleries and souvenir shops selling Russian watches, jewelry, new antiques and embroidered linen ware.

Thirty Six Streets is also the home of numerous "mini hotels," some designed for backpackers, although you increasingly find facilities which include air-conditioning, color satellite television, and international dial-direct telephones. It is also home to backpacker cafés, such as the **Darling Café** ((4) 826 9386, 33 Hang Quat Street, and **Green Bamboo** ((4) 826 8752, 42 Nha Chung Street, which tend to concentrate less on food and more on selling cheaper alternative tours to attractions such as Ha Long Bay and Sa Pa, and one day pagoda tours.

The Old Quarter is the symbol of Vietnamese entrepreneurial spirit at its best. Virtually every inch of the area has a shop or pavement vendor operating on it. Some of the best street food can be found in these tiny pavement eateries and it is not uncommon to see a big car pull up, its occupants piling out to sit down at the doll-size chairs and indulge in some delicious specialty before heading off to other ritzier entertainment.

It is an easy walk from the 36 Streets to **Hoan Kiem Lake** which sits right in the heart of the tourist hub. The lake is the subject of an old legend — now commemorated by a small pagoda in the middle of the lake — in which a golden tortoise snatched a magical sword which the famed fifteenth-century warrior, Emperor Le Loi, had used to drive the Chinese from northern Vietnam. The tortoise disappeared, returning the weapon to the gods, hence the lake's name which translates in English to **Lake of the Restored Sword**.

On the lake's northeastern shore, close to Silk Street market district, is the small and extremely picturesque **Ngoc Son**, or **Jade Mountain Temple**, reached by a traditional Chinese-style humpbacked wooden bridge called **The Huc (Touched by Morning Sunlight)**. Ngoc Son is dedicated to General Tran Hung Dao, who drove the Mongols out in the thirteenth century (and who is commemo-

rated with a towering statue in Saigon's Hero Square); but the temple's main attraction is its bridge and its location — a popular venue for photographers. Every morning, the lakeshore attracts a local enthusiasts for tai chi exercises and badminton.

Of all Hanoi's colonial buildings, the **Municipal Theatre** is the most centrally located and the most splendid. Standing at the eastern end of Trang Tien Street, a short walk from the south of Hoan Kiem, it dominates a huge square formed by six intersecting roads. Built in 1911 as an opera house, it is still more commonly referred to as **The Opera**. The Communist takeover of Hanoi on August 16, 1945, was proclaimed from one of its balconies. In 1992 it provided the opening orchestral setting of the French movie

Dien Bien Phu. Today, the opera house and its still-elegant 900-seat auditorium is the cultural center of Hanoi, packed each night for performances of anything from the Hanoi Symphony Orchestra to operatic recitals, traditional music and dance, Vietnamese pop or Western rock shows.

To the north of the 36 Streets, the huge **Dong Xuan Market** has several floors packed with electronic goods, household products, linen and merchandise of every description. An open market for vegetables, fresh fish, and produce sprawls around it — crowded beyond imagination every morning with rural vendors in their conical hats, hauling bicycles and bamboo baskets full of produce in from the country. Sadly, the government is concerned about the "messiness" of these street vendors and is in the process of chasing them off the streets. Whether they will build a new market on the outskirts of town is a matter of conjecture, but it would be another victory for development, and mean good-bye to the people's fresh vegetable supply, and to yet another facet of traditional life.

On the northern side of this melee you'll find a small Taoist temple, surrounded by gardens but hidden discreetly behind high walls, with prayer halls adorned with images of Taoist gods, Buddha and immortals.

Built in 1886, the remarkable square-towered **Saint Joseph Cathedral** on Nha Chung Street, west of Hoan Kiem Lake, is a well-weathered shell in an unkempt garden,

Hanoi's colonial-era Opera House still flourishes after two wars and a Marxist revolution.

its stone walls stained and flaked by the elements. But when the light is right, usually in the late afternoon sunshine, it becomes a medieval vision — the effect made all the more profound by weathering and neglect. It's a photographer's dream. Visitors are quite welcome to take part in Mass, which is held twice a day from 5 AM to 7 AM and from 5 PM to 7 PM. For the rest of the day it's bolted shut, and no matter who you talk to, no one seems to have the key.

The mustard-colored **Government Guesthouse** is another photographer's paragon, an ornate colonial mansion surrounded by a wrought-iron fence which was once the palace of the French governor. Located on Ngo Quyen Street opposite the **Sofitel Metropole Hotel**, it invites closer inspection — but entry is restricted to official state guests with little hope of non-VIPs talking their way beyond the locked gates. However, it stands along with the Sofitel Metropole as one of the finest examples of colonial architecture in Hanoi.

At the northeastern end of the Old Quarter, the dramatic 1,682-m (5,518-ft) **Long Bien Bridge**, which was bombed and strafed repeatedly during the Vietnam War until American prisoners of war were put to work repairing it, provides another key conduit into the district. Every morning the bridge throngs with farmers and workers from the countryside beyond the Red River, many of them struggling across with bicycles and baskets loaded with chickens, ducks, vegetables and fruit to grab a vantage point in the turmoil around Dong Xuan. The bridge is one of only two that span the Red River.

West Lake

Further to the west and lying on Hanoi's northwestern fringe, **West Lake**, or **Ho Tay**, is a tranquil recreational area best viewed at sunset, when the sky and water become a sheet of vivid, changing color. Along the lake's eastern shore, a causeway separates it from a smaller body of water, **Ho Truc Bach (White Silk Lake)**, which is also called the "bomb lake." This is where, in October 1967, an American pilot landed after bailing out of his crippled jet, which had been hit by antiaircraft fire during an attack on an electronics factory. A rather crudely sculptured

sandstone plaque at the lakeside commemorates the event.

Both West Lake and Truc Bach Lake are now more noted for the new hotels and open-air restaurants which have mushroomed around their shores. On Sundays, particularly, they're packed with wealthier Vietnamese families and are an excellent place in which to mingle and chat.

On the eastern shore of West Lake, alongside Thanh Nien Street which runs north between the two lakes, is a fifteenth-century temple which was rebuilt in 1842. **Tran Quoc Pagoda** features a stele believed to have been erected in 1639 which recounts the temple's history, and a main prayer hall and altar decorated with row upon row of Taoist and Buddhist images.

At 44 Ngu Xa Street, standing at the southern end of Thanh Nien Street on Truc Bach Lake, the Ly Dynasty (1010–1225) **Thanh Quang Pagoda** has a huge hand-cast bronze Buddha image almost four meters (13 ft) tall, the biggest bronze statue in the country, weighing almost ten tons. It was made in **Ngu Xa** village, famous for bronze casting. The Buddha and a bell date back to 1677. What makes the temple even more appealing is that its main courtyard is the setting each late afternoon for children's martial arts practice — and another opportunity to relax and chat with local people.

You'll find the ancient university, the **Temple of Literature** or **Van Mieu**, an acclaimed example of early Vietnamese architecture. It is located about two kilometers

(slightly over a mile) west of the city center at the intersection of four streets, Nguyen Thai Hoc, Hang Bot, Quoc Tu Giam and Van Mieu. Set in spacious tree-shaded gardens, most of this complex of walled courtyards and pavilions was built in 1070, dedicated to Confucius, and became Vietnam's first university six years later, educating the sons of mandarins. A number of steles, each erected on a stone tortoise, record the names of scholars who were successful in the civil service examinations held there from 1442 to 1778. One structure, the **Khue Van Pavilion**, was built as late as 1802, and the complex underwent repair and renovation in 1920 and 1956.

Hanoi's Central Market lies roughly at the borderline of the city's old and "new" quarters.

Housed in the former French Ministry of Information at 66 Nguyen Thai Hoc Street, close to the Temple of Literature, the **Fine Arts Museum** features some good examples of Vietnamese sculpture, painting, embroidery, lacquerware and other art, but the same people who deified Ho Chi Minh's remains have been at work here too, turning traditional crafts into a commentary on the Communist triumph.

Other Areas

Lying to the south of Hoan Kiem, the smaller **Thien Quang Lake** is another early morning venue. In fact, it's something of a shock to find such huge crowds in the tree-lined squares and pathways around it — and so early, just after dawn. At the southern end, off Tran Nhan Tong Street, you'll find rows of housewives bending and swaying to the instructions of tai chi instructors. Elsewhere, hundreds of children and adults take part in badminton tournaments, keep-fit classes and all sorts of other sports and pursuits, reflecting the government-imposed social discipline that's been a feature of Hanoi life since the revolution.

The delightful wooden structure of the **One Pillar Pagoda (Chua Mot Cot)**, set in a lily pond, was built by Emperor Ly Thai Tong, whose reign lasted from 1028 to 1054, to celebrate a dream in which he was presented with a son and heir by the Goddess of Mercy, Quan The Am Bo Tat. Right after that visitation, he married a young peasant girl who bore him his first son. He built the pagoda in 1049, which was destroyed by the French in 1954. This newer pagoda was rebuilt by the revolutionary government. Today, childless Vietnamese couples pay homage at the shrine, praying for a son. You'll often find young artists there, sketching it from all angles amid the trees.

The pagoda's location, off Ong Ich Khiem Street at the southern end of Ba Dinh Square, makes it easy to combine with a visit to **Ho Chi Minh's Mausoleum**. This towering, marble-clad, Soviet-style edifice radiates such power and melancholy that it's a wonder "Uncle Ho" isn't turning in his grave — or in the glass sarcophagus that holds his embalmed corpse. It's known that in his will, Ho ordered that his remains be cremated.

Nonetheless, this is still the most revered monument in Vietnam, and **Ba Dinh Square**, which sprawls before it, the site of annual Kremlin-style victory parades. The whole area is closed to traffic, and on most days the only movement you'll see is a uniformed, armed guard stretching his legs in the trees alongside the mausoleum as an endless stream of cone-hatted, devout peasants await their turn to pay their respects to their revered old leader. Visiting days are Tuesday, Wednesday, Thursday and Saturday from 7:30 AM to 11 AM, and there are obviously a

lot of rules to follow — absolutely no photography inside the building; no stopping to gaze at the revered remains; no cameras or bags admitted; no hats; no shorts or tank-tops; no hands in the pockets; no spirited or demonstrative behavior. Each year, Ho's corpse is taken to Moscow for a couple of months — usually September to November — for remedial work to keep it preserved.

Near the mausoleum is **Ho Chi Minh's official residence**, built in 1958, and the 1906 former palace of the French governor-general of Indochina, now the **Presidential Palace** and used for official receptions.

The **Hai Ba Trung Temple**, dedicated to the Trung sisters and their rebellion against the Chinese in AD 40, isn't on Hai Ba Trung

Street—it's about two kilometers (a little over a mile) to the south of the city center on Tho Laos Street. It features a statue of the two ill-fated sisters kneeling Joan d'Arc-style with raised arms. It will be recalled that after proclaiming themselves queens of the Red River Delta region, they committed suicide rather than surrender to a Chinese counterattack.

The marvelous new **Vietnam Museum of Ethnology** is located about 10 km (six miles) out of town, at Dich Vong, Cau Giay District, in the western outskirts of the city. The bright new building houses a compre-

be friends" seems not to be the issue here, rather a "lets glorify our brave deeds" attitude seems to exist in the north, which of course is completely justified.

Situated suitably on Dien Bien Phu Street, the **Army Museum** features weaponry and scale models depicting Vietnamese victories from the French collapse at Dien Bien Phu to the 1975 fall of Saigon.

At 25 Tong Dan Street, the **Revolution Museum** is no less significant as a national monument, but you really have to have a burning interest in the documentation and

hensive ethnological collection delineating cultural facets of Vietnam's 54 minority groups which divide neatly into three broad categories: the Austroasiatic family, the Sino-Tibetan family and its relative, the Tai group, and the Austronesian group. The center has received major support from France and Holland and offers excellent introductory insights into the cultural mores of some of Vietnam's ethnic minority groups.

War Museums

An ever-growing number of museums commemorating Vietnam's struggle against French rule and what is referred to by the Vietnamese as the "American War," are burgeoning across the city. "Forget the past and

faded photographic mementos of Vietnam's emancipation to enjoy it. The **Independence Museum** at 48 Hang Ngang Street is simply the house where Ho Chi Minh drew up Vietnam's Declaration of Independence, marking the break from French rule in 1945, while the "Hanoi Hilton" has a far different story. This very truncated complex is all that remains of the once sprawling prison known as the **Hoa Lo Prison**. Situated just off Hai Ba Trung Street the prison was built originally by the French and was later dubbed the Hanoi Hilton by United States prisoners of war — mostly downed pilots — incarcer-

OPPOSITE: Hanoi's Ho Chi Minh Mausoleum.
ABOVE: Reminders of the apocalypse in the Hanoi War Museum.

ated there, often after being paraded through the streets, during the Vietnam War. While most of the original site has been used to build the glittering Hanoi Tower, a part of the original complex has been opened to visitors.

Opened early in 1998, at a cost of US$1.4 million, following the relocation of 60 households, and after 10 years in the making, the **B-52 Victory Museum** houses a significant collection of B-52 parts. The exhibition of plane parts, which will eventually be welded together, is aimed at showing visitors just how big the B-52s were and how the tiny

Vietnamese defeated them, in a "David and Goliath" sort of way. As well as B-52 parts, the museum houses Ho Chi Minh's typewriter and revolutionary slogans.

Performing Arts

For all its historical and cultural monuments, Hanoi is at its most interesting when you're simply out in the streets. It offers something that is perhaps more pronounced here than in any other city in Indochina—a previously restricted, disciplined, austere society that has burst into life and is enjoying every minute of it. Evening entertainment worth a visit include the water puppet theater and the Hanoi circus. The **water puppets** stem from an old village tradition, and if you can't witness a village performance, then the next best thing is to attend the **Kim Dong Theater** ((4) 824 9494, at 57 Dinh Tien Huang opposite Hoan Kiem Lake, for an authentic show. Water Puppet performances are held every night.

The **Central Circus** (822 9277 performs nightly except Monday at the northern gate

of Lenin Park, while the **Hanoi National Opera** (826 7361, at 15 Nguyen Dinh Chieu, has shows on Mondays, Wednesdays and Fridays at 8 PM.

Hanoi Cai Luong Theatre at 72 Hang Bac Street has performances every Saturday and Sunday night for US$2. The 70-year-old theatre has achieved recognition as a classical theatre where players dressed in costumes similar to those of Chinese opera reenact tales that vary from classical to modern themes of love and deception. While performances are in Vietnamese, the color and costumes almost make a visit worthwhile.

WHERE TO STAY

When it comes to five-star accommodation and service, the hotel that has held the fort for almost ten years is the elegant 244-room **Sofitel Metropole** ((4) 826 6919 FAX (4) 826 6920, on Ngo Quyen Street (reservations can also be booked through Resinter Worldwide Reservations, see ACCOMMODATIONS, page 270 in TRAVELERS' TIPS). Formerly the Thong Ngat Hotel, it was renovated by the French group Accor to capture the fledgling but growing business travel market in Hanoi. The result is a deluxe property incorporating all its former colonial architecture and fittings, with a top-class bar and main continental restaurant, business center, swimming pool, and plans for more rooms, executive offices and conference facilities, and additional restaurants. Rates are naturally in the high end — ranging from US$229 to US$550 for a double room.

However competition has reared its head in the new **Hanoi Hilton** adjacent to the Opera, but no reservation details were available at the time of writing (see ACCOMMODATIONS, page 270 in TRAVELERS' TIPS). Apart from the Metropole, there's a crop of new hotels and guesthouses that offer a slightly less heady price bracket, if less atmosphere. One new hotel which is gaining repute is the 152-room **Guoman Hanoi Hotel** ((4) 822 2800 FAX (4) 822 2822, at 83 A Ly Thuong Kiet Street in the Central Business District. Rates are high, from US$190 to US$290, and facilities include the recommended 24-hour Paradise Café. The glitzy **Daiwoo Hotel** ((4) 831 5000 FAX (4) 831 5010, at 360 Kim Ma, offers rooms from US$199 to US$1,500. The **Horizon Hotel** ((4) 733 0808 FAX (4) 733 0888, at 40 Cat Linh has high-priced rooms from US$114 to US$240. The **Guest House of the Ministry of Defense (Nha Khach Bo Quoc Phong)** ((4) 826 5541 FAX (4) 825 9276, at 33A Pham Ngu, Laos, close to the Sofitel Metropole, is plain but secure, with rooms from US$35 to US$45.

A new hotel worth trying is "Hanoi's first boutique hotel," the 33-room **De Syloia** ((4) 828 5346 FAX (4) 824 1083, at 17A Tran Hung Dao, with moderately high-priced rooms

specialties. There are literally dozens of hotels in the Old Quarter. Another worth trying is **Stars Hotel** ((4) 828 1911 FAX (4) 828 1928, at 26 Bat Su, which has excellent service and a central location.

WHERE TO EAT

With the influx of expatriates working in Hanoi, food is abundant and diners are spoiled for choice with a range of cuisines from Thai to Mexican, not to mention French and Italian and, of course, excellent Vietnam-

from US$108 to US$198. The hotel comes with full facilities including 24-hour room service, satellite television and international dial-direct phones.

For my money, in Hanoi I choose to stay in the Old Quarter, where for US$20 to US$35 you can find a perfectly comfortable hotel with sufficient facilities. In addition, you get to walk out the door right into the center of the major tourist attraction, not unlike walking into a movie. Recommended is the cozy **Lavender Hotel** ((4) 828 6723 or (4) 828 6046 FAX (4) 828 7123, at 3 Tong Duy Tan, where a moderate US$30 single or double will secure you friendly service, air-conditioning, international dial-direct phone, satellite television, and a menu of Vietnamese

ese. For starters, the Sofitel Metropole's **Le Baileau Continental Restaurant** (see WHERE TO STAY, above) offers the most elegant setting and best cuisine in Hanoi, along the highest prices: dinner for four with drinks can set you back about US$200.

One of the most progressive and trendy restaurants is **Miro** ((4) 826 9080, at 3 Nguyen Khac Can near the Opera, which offers a varied and innovative continental menu. The **Press Club** ((4) 934 0888, at 59A Ly Thai To Street, has a delightful deli below and fine dining upstairs. A good Italian restaurant to try is **La Primavera Ristorante Romano**

OPPOSITE: The Sofitel Metropole remains one of Hanoi's most prestigious hotels. Street barber ABOVE bears scars of the Vietnam War.

((4) 826 3202, at 12 Pho Hue. Also recommended are **A Little Italian** ((4) 825 8167, at 78 Tho Nhuom; the small and cozy timber interior of **Mediterrano** ((4) 826 6288, at 23 Nha Tho; **Il Grillo** ((4) 822 7720, at 116 Ba Trieu; and **Al Lago** ((4) 718 4027, at 5 Xuan Dieu, which has an Italian chef. For a pricey Italian meal, try the fancy **La Paix** ((4) 831 5000 extension 3245. 3245, at Daiwoo Hotel 360 Kim Ma.

Thai food is, not surprisingly, also popular. **Siam Corner** ((04) 258 1200, near Westlake in the Oriental Park, Quang An, Tay Ho

District close to the Tay Ho Pagoda, is reputed to be the best in Hanoi (combining agendas could be a good idea here — a temple visit and lunch). Another to try is **Baan Thai** ((4) 828 0926 at 3B Cha Ca Street (near the famous fish restaurant described below), where dishes go for US$3 to US$6. A mixture of Asian food, notably Singapore and Malaysian classics as well as Asian fusion creations, can be enjoyed at the highly recommended and moderately priced **Café Paradise** ((4) 822 2800 at the Guoman Hotel at 83A Ly Thuong Kiet Street, where delicious dishes come at surprisingly reasonable prices with a full meal costing around US$10.

Hanoi street vendor.

Vietnamese food comes in abundance, and listed below are several restaurants located in converted villas. Recommended are **Nam Phuong** ((4) 824 0926 at 19 Phan Chu Trinh, Hoan Kiem District (one American couple I spoke with ate here every night); the **Bon Mua** (Seasons of Hanoi) ((4) 843 5444 at 95B Quan Thanh Street; and **Com Nieu Que Huong** ((4) 851 7140, at 250 Ton Duc Thang. **Indochine** ((4) 824 6097, 16 Nam Ngu, comes with plenty of atmosphere, a marvelous name, but perhaps not the most exquisite food. You can find classic Vietnamese cuisine at **Cay Cau** ((4) 824 5346, 17A Tran Hung Dao Street, with indoor and outdoor dining; and try **Van Xuan** ((4) 927 2888, 15A Hang Cot, for excellent Hue food. **Hollywood Café** ((4) 943 1343, 10 Ho Xuan Huong, serves Vietnamese food for breakfast lunch and dinner in a Vietnamese new-wave decor with singers and dancers.

French cuisine can be enjoyed at **Gustave Restaurant Français** ((4) 825 0625, 17 Trang Tien, or **Lan Anh Restaurant** ((4) 826 7552, at 9A Da Tuong, which does French, Algerian and Vietnamese food, specializing in couscous. **Le Splendide** ((4) 826 6087 serves traditional southwest French cuisine at 44 Ngo Quyen, and **Le Cyclo** ((4) 828 6844, at 38 Duong Thanh, serves Vietnamese and French cuisine in cyclos, or the garden behind offers table-barbecued specialties.

Not-to-be-missed Hanoi institutions which immediately evoke the old French-Vietnamese atmosphere include the very atmospheric **Cha Ca La Vong** ((4) 825 3929 at 14 Cha Ca Street in the Old Quarter, or its newer incarnation at 107 Nguyen Truong To Street (((4) 823 9875), where they have served the same delicious herbed, fried fish dish for three generations, all washed down with cold beer. They are open daily from 10 AM until 10 PM. My advice is be adventurous, even small family run places can be surprisingly good.

HOW TO GET THERE

Hanoi is accessible by around ten international and regional flights as well as a strong domestic Vietnam Airlines network. The main office for Vietnam Airlines ((4) 825 0888 FAX (4) 824 8989 is in the Hoan Kiem District

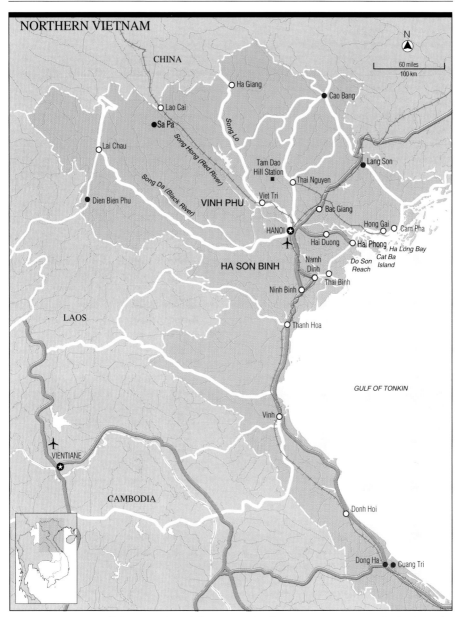

NORTHERN VIETNAM

CHINA

Ha Giang

Cao Bang

Lao Cai

Sa Pa

Song Lo

Lai Chau

Song Hong (Red River)

Lang Son

Tam Dao
Hill Station

Thai Nguyen

Song Da (Black River)

VINH PHU

Viet Tri

Dien Bien Phu

Bac Giang

HANOI

Hong Gai   Cam Pha

Hai Duong   Hai Phong   Ha Long Bay

HA SON BINH

Namh
Dinh   Do Son   Cat Ba
Reach   Island

Thai Binh

Ninh Binh

LAOS

Thanh Hoa

GULF OF TONKIN

Vinh

VIENTIANE

CAMBODIA

Donh Hoi

Dong Ha   Cuang Tri

N

60 miles
100 km

at 1 Quang Trung. All the major hotels have a pickup service, and if you're not being picked up, it is far better to take the Vietnam Airlines minibus for US$4 than a taxi which will cost around US$25. The mini bus will drop you off outside your hotel.

ENVIRONS

The ancient lands around Hanoi are dotted with dozens of venerable Buddhist temples and pilgrimage sites, many of which can be visited with the minimum of fuss on several day tours arranged from Hanoi. As an alternative, taking your own transport would allow more freedom.

Chua Tram Gian, about 25 km (17 miles) along Highway 6 is the first pagoda to reach. This twelfth-century pagoda, founded in 1168 and restored in the seventeenth and eighteenth centuries looks out from a hill across a landscape of rice fields.

The **Tay Phuong Pagoda** (West Pagoda), situated about 40 km (27 miles) southwest of Hanoi, is a complex of three small pagodas housing 79 lacquered wooden carvings. Dating from the eighth century, these carvings illustrate stories from Buddhist scriptures.

The nearby eleventh-century **Thay Pagoda** (Master's Pagoda) was built into the side of limestone cliffs rising from the rice fields. It is dedicated to Thich Ca Buddha and 18 *arhats*, or monks that have gained Nirvana. A statue of the eleventh-century monk, Tu Dao Hanh — the Master after whom the pagoda was named — is placed near the central altar. Within this beautiful complex is a small pond and the Dinh Thuy, a small temple dedicated to one of Vietnam's unique performing arts — water puppetry — where performances are staged for pilgrims and visitors at the temple's annual festival from the fifth to seventh day of the third lunar month (March/April). These two pagodas can easily be visited in one day.

The **Chua Huong Pagoda**, or Perfume Pagoda, is one of the most popular day trips from Hanoi, about 60 km (40 miles) southwest of the city in Ha Son Binh Province. This complex of pagodas and Buddhist shrines built into the limestone cliffs of Huong Tich Mountain is regarded by many as Vietnam's most beautiful spot. A steady stream of small wooden rowboats transport visitors for an hour through verdant rice fields, dotted with pagodas and surrounded by towering and often misty limestone karsts, passing several smaller pagodas before reaching the main pagoda. It is especially busy during the pilgrimage season, which runs from February to April.

Hoa Lu, the ancient capital of northern Vietnam, which flourished in the Dinh and early Le Dynastys between 968 and 1009, lies in a region of karst hills. It is often described, for the benefit of tourists, as an inland Ha Long Bay, 95 km (63 miles) south of Hanoi on Route 1, not far from Ninh Binh. Though once rivaling Hue as a royal citadel, and covering an area of three square kilometers (just over one square mile), all that's now left of this once grand city are a shrine to Confucius and two sanctuaries, restored in the seventeenth century, commemorating the Le emperors — one

of them featuring early weaponry, drums and gongs amongst other relics.

Built by the French in 1907, the hilltop settlement of **Tam Dao** is northern Vietnam's version of the south's Da Lat — a highland retreat where the colonials sought relief during the hottest summer months. Some of their villas are still there, though suffering considerably from years of neglect, but the environment has remained fairly pristine — a series of wooded peaks offer hiking amid giant ferns and wild orchids, and visits to hill tribe communities. Tam Dao is in Vinh Phu Province about 85 km (53 miles) northwest of Hanoi.

The large natural reserve of **Cuc Phuong National Park** was established in 1962 — a hilly rainforest embracing a wide variety of flora, insects, animals and reptiles. But it's a long way from Hanoi — 140 km (87 miles) by road, with a lot of potholes and washed-out sections along the way. However, the park has a guesthouse at its headquarters for overnight visits.

The riverside border town of **Lang Son**, 150 km (93 miles) north of Hanoi, was the last stop in Vietnam on the railway to Nanning, China, when the link was operating. It sprang back into prominence in the 1979 border war, when it was almost destroyed by invading Chinese forces. In the reconciliation that followed, Lang Son has become one of the major conduits of trade — official and unofficial — between the two former enemies; real backwoods Vietnam and interesting as it is, it's right off the tourist map. The Lang Son district is home to several ethnic minority groups which include the Dao and Nung.

HAI PHONG

This rather dilapidated coastal city, 103 km (64 miles) east of Hanoi, is Vietnam's third largest city, and is strategically important as the key port of northern Vietnam. From a visitor's point of view, it is important too as the gateway to the fishing port of Hong Gai and the Cat Ba Islands, with ferries leaving from the main city terminal, a short cyclo ride

Municipal Theatre TOP in Hai Phong. Decrepit vehicular ferries BOTTOM provide the only means of crossing Hai Phong's Cam River.

from Hai Phong Railway Station. It also provides the main road access to Ha Long Bay, northern Vietnam's dramatic vacation "resort," and the route north to this beautiful bay is even prettier than the approach from Hanoi — taking two river ferries and running through farming communities before melting into an almost surrealistic landscape of green fields and towering karst formations.

Although a prime United States bombing target during the Vietnam War, the regular presence of Soviet freighters — and the risk of a much more dangerous conflict —

while waiting for the ferry to Hong Gai or to the Cat Ba Islands, it is worth catching a cab to see the **Du Hang Pagoda** located to the south of the city center at 121 Du Hang Street. This tenth-century pagoda, rebuilt in 1672 is the oldest in the city and of some architectural interest.

The **Hang Kenh Tapestry Factory** and the **Hang Kenh Communal House** on Hang Kenh Street, to the southwest of the downtown area, are worth a stop. The communal house features a collection of 500 wooden sculptures, and the 65-year-old Hang Kenh

saved it from destruction. Nonetheless, it was mined by the Americans to halt the flow of Soviet military supplies. Military aspects aside, Hai Phong is the proud possessor of some wonderful, although dilapidated French colonial buildings lining the river which in the old days was the main harbor.

WHAT TO SEE AND DO

Hai Phong is a bustling commercial city and market place; much of its character comes from being an international transhipment center. Its shops and open markets offer consumer goods straight off the ships. While Hai Phong is not blessed with huge numbers of attractions, if one has an hour or two to spare

Tapestry Factory produces traditional and modern woolen carpets and tapestries for export.

Twenty-one kilometers (14 miles) southeast of Hai Phong is the hilly coastal peninsula and beach of **Do Son**, once a popular seaside resort for the French. Crumbling colonial villas still line the shore. Do Son is now best known for its casino.

WHERE TO STAY

Being an international port, Hai Phong has a number of hotels that cater to businessmen. The best place to stay is Hotel du Commerce, which is now called the **Huu Nghi (** (31) 842 706, 62 Dien Bien Phu Street. This well renovated place of faded French elegance, operated by Hai Phong Tourism, offers moderately priced rooms from US$35 to US$40. Alternatively it's possible to stay on the beach at the **Hai Au Hotel** ((31) 861 272 FAX (31) 861 176, on Do Son Beach next to the casino.

The **Hoa Binh Hotel** ((31) 859 029, 104 Luong Khanh Thien, is popular with back-packers. A new annex has sent the rates and quality soaring.

WHERE TO EAT

There are a number of small, friendly restaurants and coffee shops which have sprung up around the central market area of Hai Phong, most of them offering very palatable Vietnamese food. The **Bong Sen Restaurant** at 15 Nguyen Duc Canh Street is said to be "the best Western and Vietnamese restaurant in Hai Phong City."

HOW TO GET THERE

Whether traveling by road or rail, the trip from Hanoi to Hai Phong is a major attraction, a trip which passes through some of the prettiest and most idyllic rice lands of the Red River Delta. It's along this route that you'll see Vietnam's still-traditional farming methods in action — fat water buffaloes plowing the flooded paddies; women in conical hats sowing seed and fertilizer by hand; flocks of ducks being shepherded to new feeding grounds; rows of bamboo hats nodding and bending amid the green and golden rice as the womenfolk harvest it with small curved knives. Because of the speed with which the day heats up, the farmers and their families are usually working in the fields at first light and are on their way home to shelter and rest by mid-morning. In the late afternoon, the road becomes a social spot for families gathering to chat and enjoy the cooling air.

Along the route you'll see towns and villages set far back from the road. If you're traveling individually, it's possible to arrange with your driver to make a detour here and there. If you're with a group, forget it: like much of Asia, the average Vietnamese driver or tour guide has one responsibility in mind

— to get you as fast as possible to wherever you're supposed to go. Tours pass through Hai Phong en route to Ha Long Bay.

HA LONG BAY

Until recently, to arrive in Ha Long Bay, 60 km (37 miles) north of Hai Phong, was to take a step right back in time. But with its increase in tourism, this once rustic bay is moving headlong into the 21st century. The tattered bat-wing sails of old sailing junks have virtually disappeared, and the only sails you are likely to see dotting the clear emerald waters today are those on tourist boats. Occasionally groups of canoeists, part of a different kind of tour, glide past against the dramatic backdrop of some 3,000 chalk and limestone formations that form a dragon's tail across the bay, (see SPORTING SPREE, page 32).

Some of the islands themselves have huge caves and grottoes in them — some of them extending up to two kilometers (one mile) into the rock — but whether you'll really be able to enjoy the stalactites and other formations within them depends on the efficiency of your guides: they're generally helpful and enthusiastic, but their smoky kerosene lamps don't really throw much light around you.

Ha Long is a physical attraction certainly worth seeing now, before the developers get their hands on it.

An early morning ferry crossing to Hong Gai brings you to Ha Long Bay's key port. Worth a trip on its own, this bustling working village is just about as authentic and colorful as you can get. Most of the fishing junks are based here and there is a splendid bustling fish and seafood market that starts in the early hours. The food at the market is exceptionally good.

Fast boats and slow ferries run from the harbor to the islands of the Cat Ba Archipelago, 80 km (50 miles) to the east (See NATIONAL PARKS, page 28 in THE GREAT OUT-DOORS).

OPPOSITE: Fishing family in Ha Long Bay — this region spawned much of the refugee migration to Hong Kong. OVERLEAF: Fantastical karst outcrops, island caves, and grottoes promise a rich tourism future for northern Vietnam's Ha Long Bay.

WHERE TO STAY AND EAT

The town of Bay Chai is the tourist center of Ha Long Bay, while to the north is the delightful fishing town of Hong Gai, linked by a clanking old vehicular ferry. Bay Chai, set on a long pebbled beach, is where the tourist hotels are. Most of them charge about the same rate for foreigners US$25 to US$45 depending on the floor you're on and whether you want air-conditioning or ceiling fans. Rooms are clean, and most have open balconies with

a sea view, but the bathrooms are often antique. If you are not on a tour, the hotel managers can arrange an excursion through the karst islands. The **Ha Long Hotel** ((33) 846 320 FAX (33) 846 318, at Vuon Dao, Bai Chai, is a renovated old colonial hotel extended to 110 rooms with a seafood restaurant, and **Hoang Long** ((33) 846 234 or (33) 846 318, at Bai Chai, has 56 rooms and was recently modernized. Eat at the hotels, or on the tourist strip facing Bai Chai Beach at one of the popular newer seafood restaurants.

HOW TO GET THERE

Ha Long Bay is accessible by both road from Hai Phong and by ferry from the thriving fishing port of Hong Gai, a short drive away. The bus trip can take more than three hours from Hanoi or two hours from Hai Phong, allowing for photo stops, passing by picturesque inland ports fringed by karst hills amid the rural setting. A far more pleasant route is to take the two-hour train from Hanoi to Hai Phong, then take the two-hour bus to Ha Long Bay (Bai Chai). Or take the three- to four-hour ferry to Cat Ba and use it as your base, taking advantage of one of the numerous tours from Cat Ba to Ha Long Bay run by local fishermen.

Alternatively, take the ferry to Hong Gai (from Hai Phong), then take a bus, motorbike or taxi to Ha Long Bay, only an hour including the car ferry ride from Hong Gai. There is also an expensive heli-jet operating directly from Hanoi to Ha Long Bay.

Since access is a little complicated, most visitors take a tour. Numerous tours are available from Hanoi, including many reasonably-priced tours from the travelers' cafés, as well as the larger travel agents. Of the cheaper tour companies, **Green Bamboo** ((4) 826 8752, 42 Nha Chung Street in Hanoi, is reputed to be the best. For a very reasonable US$29 they offer a two-day tour which includes two five-hour boat trips, overnight accommodation in a hotel and all meals (drinks not included). As long as you don't mind being crammed into the nether regions of a minibus, tours offer convenient two and three day packages.

MAI CHAU

About 120 km (75 miles) from Hanoi, Mai Chau is in the midst of White T'ai Territory, a panoramic and fertile valley surrounded by towering limestone karsts. Between the luscious green ricefields are sturdy villages — now a popular destination for overnight tours from Hanoi. So far the tours remain pleasant and not too tacky. If you would prefer to go on a quieter night remember that all groups leave from Hanoi on Monday, Wednesday and Saturday. The other nights — Tuesday, Thursday, Friday, and Saturday, are relatively quiet. Traditional T'ai dances are generally presented for the groups, however, which is an enjoyable night that is not to be missed. Tours are easily arranged from the tourist cafes in Hanoi's Old Quarter.

DIEN BIEN PHU

Vietnam's northern mountain ranges are home to numerous ethnic mountain minority people, and magnificent panoramas. Right on the northwestern frontier bordering Laos is the historic market town of Dien Ben Phu, located on Highway 42 — a diversion from the main route, Highway 6. Ardent mountain goers can only pray that one day foreigners as well as locals will be able to cross at the highland border posts up here. For now, foreigners have to make two arduous trips if they want to explore this magnificent and little-populated region: once in Vietnam and again in Laos.

The long drive to Dien Bien Phu is rewarded with abundant nature, a host of minority people and a lively market. In addition, there is plenty of war history for those who want to revisit the site of some of Vietnam's most heroic fighting against the French. Just 16 km (10 miles) from the border with Laos and 420 km (260 miles) northwest of Hanoi, Dien Bien Phu is part of the northwestern mountain circuit for keen travelers. Several small hotels are available.

LAI CHAU

North of Dien Bien Phu, and around 460 km (300 miles) from Hanoi along Highway 6, over the 1,016 m (3,300 ft) Din Bin (heaven and earth) Pass, Lai Chau is really out there — truly somewhere between heaven and earth. Lai Chau is part of the great northwestern road loop that runs from Hanoi west to Lai Chau and Dien Bien Phu before turning east and looping down southwards to Sa Pa and Lao Cai and eventually back to Hanoi, passing through magnificent rural panoramas. Lai Chau Province borders both Laos and China, one of the remotest parts of the country. Almost 25 ethnic groups inhabit these rugged mountains where small communities nestle surrounded by jungle. While Highway 6 sounds a comforting name, the reality consists of steep climbs, dramatic switchbacks, and abundant use of low gears — to say nothing of spectacular, sweeping panoramas. Only four-wheel drive vehicles or serious motorbikes need try to navigate it. From Lai Chau, the Highway changes its name, but not its condition, to Highway 12 as it heads north to Phong Tho then east to the relative urbanity of Sa Pa.

SA PA

Far more popular and more easily accessible is the hill town of Sa Pa. Not only is it approachable by the long mountain route via Dien Bien Phu, but it can also be reached by overnight train from Hanoi to Lao Cai, followed by a 45-minute bus ride through lush mountains.

Before the Chinese invasion of 1979, this onetime French retreat had over 200 villas. Today most of the remaining 11 villas have been somewhat renovated, but in a rather charmless Vietnamese fashion — no foreigners are allowed to get their hands on them.

Sa Pa (Vietnam) is the center of several tribal groups who converge in town on Saturdays for the morning market, and at one time for the Red Zao courtship practice, where the lads would sing poetic stanzas to girls dressed in all their tribal finery. They have since moved elsewhere — 35 km (22 miles) away from the prying eyes and taunts of local tourists. Now the market in its new concrete building is a lively spot all week and although much of the charm has gone, it is still a magnificent place to visit.

Close to Vietnam's highest peak, the 3,142-m (10,309-ft) **Mount Fansipan** (Fang Xi Pan), Sa Pa offers marvelous trekking country in the Hoang Liem Nature Reserve, and a cool climate conducive to physical exercise (see EXPLORE THE TONKINESE ALPS, page 15 in TOP SPOTS).

Among several not-so-appealing larger hotels, the new French **Victoria Hotel (** (20) 871 522 FAX (20) 871 539 stands out with three-star comfort, quiet good taste, and a decent menu, all welcome additions to the somewhat simple (or rather, kitsch) accommodation previously available. Room prices range from US$55 to US$100. The **Post Office Guest**

The Sunday market fills with Hmong "Flower" girls at Bac Ha.

House ((20) 871 244 FAX (20) 871 282, next to the post office, also offers reasonably comfortable accommodations. Right at the bottom of the main road through town is the very popular **Dang Trung Auberge** ((20) 871 243 FAX (20) 871 282, whose delightful owner Mr. Trung speaks French and takes great pride in showing favored guests his magnificent flower garden. Rooms—some with fire places — range from US$10 to US$20. They can organize tours and a car, and the restaurant, while not serving grand cuisine, is full of red-cheeked, happy eaters. The

rade their colorful costumes while shopping for basic necessities and, for the young girls, a husband.

HEADING SOUTH

The long strip from Hanoi to Hue is often bypassed by tourists. Buses make the journey overnight as does the train, and the few places of interest along the way are barely worth the extra effort involved. The area around **Ninh Binh** with its pagodas and weaving villages can be explored as a

Telecommunication Guest House ((20) 871 398 FAX (20) 871 332 has rooms from US$20 to US$30. Those on a tight budget will find the **Rose Guesthouse**, right on Sa Pa's main street, cheap and cheerful with clean rooms from US$3.

BAC HA

An even newer destination, available through tours from both Hanoi and Sa Pa, is the Sunday market of Bac Ha, a three-hour drive from Sa Pa or two hours from the railhead at Lao Cai on the Chinese border. Here the predominant minority group are the variegated Hmong, who come from miles around to preen and pa-

day trip from Hanoi. Further to the south, Thanh Hoa, and the village of **Dong Son**, the reputed site of the Bronze-Age Dong Son culture whose influence spread through southeast Asia, has little to show of its once-great culture. Note, however, that Dong Son designs are still apparent in Lao textiles and extend to the far eastern Indonesian islands, where designs are incorporated in the weaving patterns and the distinctive Dong Son drums are still a part of the bride price.

Vinh is really only of interest as a stopover if traveling south by road. One of the poorest and more culturally backward areas of Vietnam, the region is plagued with floods and typhoons as well as poor soil. The city was bombed by the French in the 1950s and later by the Americans from 1964 to 1972. The town was later rebuilt with East German assistance, a dismal architectural feat of gray, formless concrete drabness. One site of interest is the **Kim Lien Village**, the birthplace of Ho Chi Minh, 12 km (nine miles) west of Vinh. Outside the urban areas, towards the west, the mountains of Nghe Tinh Province are heavily forested, home to numerous minority groups which include Muong, T'ai and Khmer. It is said that the mountains are still home to a wildlife population that in-

nam War it was the scene of intense fighting between United States Marines and North Vietnamese regulars in the 1968 Tet Offensive that almost completely demolished the once grand city. At that time more than half the population of Hue lived within its outer walls, and in the battle to drive out the Viet Cong, a great deal of this national treasure was destroyed.

Like Da Nang, Hue suffered considerably during the war. During the Tet occupation by Communist troops, some 3,000 pro-government people and Buddhist monks were

cludes elephants, gibbons, leopards, deer, rhinoceros, squirrels, and monkeys.

HUE

Misty romantic Hue has long been the home of poets and painters, inspired, no doubt, by its incredible natural beauty, which all the depredations of the war failed to erase.

Established around 1687, Hue was later the capital of Vietnam under the Nguyen dynasty, the last to rule Vietnam, from 1802 to 1945. During that time, the Citadel, the court of the Nguyen emperors, was almost comparable in grandeur to the Forbidden City in Beijing. From 1945, the Citadel gradually fell into disrepair, and during the Viet-

executed and buried in graves all over the city. In 1975, Hue was one of the first major south Vietnamese cities to fall to the Communists, triggering a violent mass panic in which thousands of civilians and fleeing government troops fought their way through the surrounding hills and down the coast to Da Nang. However Hue remains one of Vietnam's most charming cities, and the young women are reputed to be the most beautiful in the country.

As a tourist location, Hue is worth a visit not so much to see what's left of the Cita-

OPPOSITE: Bac Ha sugar cane merchant does a roaring trade with minority customers. ABOVE: The Hue Citadel is Vietnam's equivalent of China's Forbidden City. OVERLEAF: Sunrise on Hue's busy Perfume River.

del, but to wander through the historic and decorative tombs while enjoying the scenic and clean Perfume River, which flows right through the city and to see the marvelous Thien Mu Pagoda in the late afternoon with its views across the river. A number of Sampan communities live along the inner city riverfront and its not unusual to see the hoop-covered country boats making their way up and down the river. A lively market and small home lacquerware industries, all conspire to offer a pleasant few days of sightseeing.

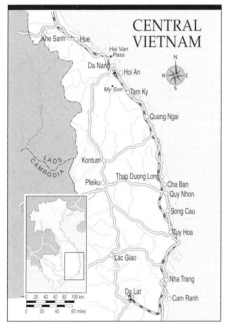

Sights aside, simply roaming through the lush Hue countryside is a pleasure in itself. On a motorbike (the traffic is minimal on the country roads), or by bicycle, it's wonderful to spend a day or two exploring the surrounding villages at your own pace, coming up against delightful surprises, like a farmer plowing his rice fields with his buffalo, while an old forgotten royal tomb stands behind as a decorative backdrop.

GENERAL INFORMATION

Hue City Tourism ((054) 823 577 or (054) 823 406 is located at 1 Truong Dinh Street, while the Hue **Tourist Office** ((054) 822 369 FAX (054) 824 806 is at 15 Le Loi Street. In

addition, small private companies offer tours to the tombs and to the Demilitarized Zone (DMZ). **Huetour** ((054) 825 242 can issue train, bus and boat tickets.

WHAT TO SEE AND DO

Although only a shadow of what it once was, the **Citadel** stands behind high stone ramparts and huge Ngo Mon (Meridian) Gates, the main access to the imperial enclosure. Looking at its sweeping yellow-tiled roofs, red wooden pillars and huge yellow doors, you're reminded of the Meridian Gate of the Forbidden City in Beijing, though it's not quite as majestic. Along the Citadel's outer walls stand other surviving ceremonial gates. Mounted with intricate carvings, colored porcelain and mirrored glass chips, and snarling dragon decorations, they give a sense of the imperial city's former scale and authority.

Sadly, the Citadel you'll visit now is really just the **Purple Forbidden City**, once the inner, exclusive domain of the emperors, where the Ngo Mon Gate leads to dank lotus pools; the vast, wing-roofed **Thai Hoa (Supreme Harmony) Palace**, which was used for state and ceremonial events; the **Halls of the Mandarins**; the recently renovated **Imperial Library**, featuring ornate ceramic sculptures of mandarins and other figures on its roof; and the nearby **Royal Theater**, which is now the home of the **National Conservatory of Music**. You're quite welcome to enter the conservatory grounds and listen to the students practicing — and catch an impromptu chamber recital if you're lucky. The **Imperial Museum**, just beyond the north wall, completes the Citadel tour — housed in another impressive hall and featuring surviving costumes, furniture, porcelain, a sedan chair, musical instruments and other relics of the imperial reign.

is particularly pleasant. Inside an octagonal stone wall, the monarch's sepulcher lies alongside a small lake surrounded by a pavilion, a temple, an honor courtyard lined with stone sculptures of elephants, horses and mandarins, and a stele pavilion sheltering a 20-ton stone tablet recording Tu Duc's virtues and triumphs.

The **Tomb of Khai Dinh**, who ruled from 1916 to 1925, reflects how Westernized even the imperial court had become by then — the architecture and statues have vague European characteristics. The stele pavilion

The evocative ruins of Hue's imperial past are dotted amongst small farms and rice fields along the Perfume River. While tours will try to cover all the tombs, be selective and choose just the best, allowing more time to enjoy the peaceful surroundings. To visit the tombs, join a group tour or rent your own covered sampan (check with your hotel or try at the boat depot adjacent to the Century Hotel where boats are available for US$30 to US$40 for the day), and proceed at a relaxed pace.

Four locations inspire real interest. The **Tomb of Tu Duc**, who ruled from 1848 to 1883, is among the closest, located in **Duong Xuan Thuong Village** about seven kilometers (just over four miles) from town. It's best to visit in the afternoon, when the light

is in the middle of the traditional honor courtyard, with its guardian beasts and mandarins, and an image of Khai Dinh himself sits under a concrete canopy in the complex's ornate main hall, its walls decorated with elaborate frescoes.

The most impressive tomb, the tomb of **Emperor Minh Mang** (1820–1840), lies 12 km (seven and a half miles) from Hue at **An Bang**, one kilometer (a little over half a mile) from a beautiful section of the Perfume River. Three great ceremonial gates and three granite staircases lead to the **Stele Pavilion**. Three terraces then lead to **Sung An Temple**, dedicated to the emperor and his empress. Three bridges

On Hue's Perfume River sampans are still a common means of transport.

cross the tiny **Lake of Impeccable Clarity**, the central one made of marble and used only by the emperor. Finally, three more terraces representing Heaven, Earth and Water provide the foundation for another magnificent building, the **Minh Lau Pavilion**. Minh Mang's burial mound lies inside a circular wall representing the sun at the top of a stone staircase flanked by sculptured dragons. To finish off the day, make sure your boat reaches the **Thien Mu Pagoda** in the late afternoon to watch the sun setting across the Perfume River.

There is no other place of worship in Vietnam in such a beautiful location as this renowned Thien Mu Pagoda, a seven-story octagonal tower and adjacent prayer hall perched on a promontory right over one of the most dramatic sections of the Perfume River. The view is wonderful — the river widening and stretching through green pastures and paddies toward misty blue mountains. The pagoda was founded in 1601, but the present tower was built by Emperor Thieu Tri in 1844. Adjacent to the tower, in a two-story pavilion, there's a giant bell which was cast in 1710 and which the monks will toll for you if you ask them. They say you can hear it resonate up to 10 km (six miles) away. There's also an Austin car that was used to take the bonze Thich Quang to Saigon in 1963 — where he was one of the Buddhist martyrs who burned himself to death in protest against the Diem regime. You'll come across several temples in Hue, but this is the most appealing.

Back in Hue central, on the right bank of the river, in Nguyen Hue Street, the strange spire of **Notre Dame Cathedral** rises up over the skyline, creating a point of reference from many vantage points. The cathedral is a grand but somewhat bizarre blend of European and Vietnamese architecture, and the spire is distinctly Asian. Actually, Notre Dame Cathedral was built between 1959 and 1962, but what it lacks in history it certainly makes up for in appearance.

WHERE TO STAY

Most tour groups are taken straight to the **Huong Giang Hotel** ((54) 822 122 FAX (54) 823 102 which is located right on the banks

of the Perfume River at 51 Le Loi Street, facing across to Hue's distant central market. It's a rather cutely kitsch Vietnamese hotel, with a lobby full of backbreaking traditional rosewood furniture and lots of lacquer. The panoramic view of the river and bustling sampan communities compensate for a lot and the ground-floor restaurant terrace is a fine place to sit and watch the river traffic. Rates are from US$50 to US$230. The hotel operates a villa at 3 Huong Vong called **Indochine** ((54) 826 070 FAX (54) 826 074, and the upstairs rooms are nice with rates at around US$45.

Right next door to the Huong Giang is the monumental **Century Riverside** ((54) 823 390 FAX (54) 823 394 which is also popular with tour groups but has less atmosphere.

Rooms range from US$65 to US$150. The renovated **Morin Hotel** ((54) 823 526 FAX (54) 825 155 at 30 Le Loi Street was once a backpackers' favorite and is now a very attractive place with a giant garden courtyard and a small pool. The very large and comfortable rooms with balconies cost from US$45 to US$85. The **Thang Long Hotel** ((54) 826 462 or (54) 826 463 FAX (54) 826 464 at 16 Hung Vuong Street has large comfortable rooms with a price range, from US$10 to US$60, that seems to cater to a wide audience, almost all of the offers being good values.

WHERE TO EAT

The **Huong Giang Hotel** ((54) 822 122, 51 Le Loi Street, has the Hoa Mai and Royal Restaurant serving imperial Vietnamese as well as Chinese food.

Dine on the river at the **Song Huong Floating Restaurant** ((54) 826 655 on Le Loi Street near the bridge, where seafood and "steamboat" — a variety of meats or seafood and vegetables cooked on the table in a rich meat stock — are specialties. You can eat in the Citadel at the **Huong Sen Restaurant**, a pavilion-style establishment set over a lotus pond. The **Am Phu** ((54) 825 259, 35 Nguyen Thai Hoc, serves Hue specialties in a simple setting. The newly renovated **Morin Hotel** ((54) 823 039 has a marvelous tree-shaded garden courtyard — perfect for lunch or dinner. The grounds of the **Thang Long Hotel**

The Linh Mu Pagoda in Hue, overlooking the Perfume River.

have a popular, everyday restaurant facing Hung Vuong Street which serves inexpensive food in a pleasantly bustling atmosphere.

HOW TO GET THERE

Vietnam Airlines operates regular services from Saigon, Hanoi and Da Nang to Hue's Phu Bai Airport, which is situated about 17 km (just over 10 miles) south of the city.

If you are coming from the north, the overnight train from Hanoi is a fine way to arrive. Book ahead and reserve a soft sleeper

for optimum comfort at around US$55. Leaving Hanoi around 9 PM, it arrives at Hue around the following noon, after a pleasant morning traveling through the beautiful countryside.

If you're coming from Da Nang, 108 km (67 miles) to the south, both the road and train trip are extremely enjoyable. The trip takes about three hours — six if you allow for photo-stops and lunch — and it takes you through some of Vietnam's most scenic countryside. The road switchbacks through three high mountain passes interspersed with coastal farmland, fishing communities and a couple of vast agricultural plains set around wide bays. Once in Hue, you should be able to rent bicycles, or better yet a motor-

bike, for a nominal sum. Check at your hotel, they may rent them themselves, or they will be able to recommend a place that does.

HAI VAN PASS

This 496-m (1,627-ft) spur in the towering barrier of the Truong Son Mountains, 30 km (19 miles) north of Da Nang, is the first and highest pass on the route of National Highway 1 to Hue. It marks the great gulf between north and south, a traditional barrier which the Chinese invaders never managed to conquer. It is not only the climate that seems to change at this latitude, the people too become sunnier and more relaxed, part of the legacy of the south.

Both sides of the pass enjoy spectacular panoramas — to the south, it looks over Da Nang and its coastal headlands and bays, while to the north it looks across a valley to the coastal plains on the way to Hue. In the more immediate vicinity, it's interesting to watch the dilapidated long-distance "express" buses that labor to and from Saigon and Hanoi, absolutely packed to the roof with passengers and luggage, hauling themselves at a painful snail's pace up both sides of the mountain. Everyone stops at the Pass. While passengers scramble out of their crowded coaches to stretch their legs and to photograph the spectacular view, enthusiastic hawkers descend with brimming baskets and a scrimmage begins, the hawkers usually the victors, but never mind — it provides great amusement for other coach-loads who have already survived it.

Hai Van Pass also looks down on two of Da Nang's longest beaches, **Nam O** to the south and **Lang Co** to the north — neat palm-fringed crescents of sand which are still relatively undeveloped.

DA NANG

The approach by air to this key port of central Vietnam reveals a city surrounded by beautiful beaches, bays and hills. It also reveals the pockmarks and the bomb scrapes of Da Nang's suffering during the war, and

OPPOSITE: Minh Manh Tomb in the imperial burial grounds outside Hue. ABOVE: Sampan and harvested bamboo on the Perfume River.

how the ever-innovative Vietnamese farmers have used them to advantage, converting them to fish ponds in the midst of their *padi* fields.

As a major United States Marine base, the conduit to both the Central Highlands and base camps up toward the DMZ (Demilitarized Zone), the city was the scene of some of the fiercest fighting of the war. When it fell to the advancing Communists in 1975, it became a nightmare of violence as thousands of South Vietnamese soldiers and civilian refugees fought to escape by sea to ports further south.

Due no doubt to negative after effects of the war, Da Nang continues to exude an edge of nastiness — belligerent children and beggars crowd around you the moment you step into the streets, and several incidents of major theft from hotel rooms have been reported (I had an expensive shotgun microphone stolen from mine). This is no doubt why many people give Da Nang a miss, except for Vietnam war veterans and those on tour packages. Others stop to visit the important **Cham Museum** before heading on to Hue to the north or to the charming ancient trading port of Hoi An just to the south.

Da Nang is well located for future development. Lying between the Han River and the South China Sea, flanked to the north and

south by beautiful white-sand beaches that seem to go on forever. Only the **Furama Beach Resort Da Nang** has capitalized on this beach until now.

To its north and west, the foothills of the **Central Highlands** begin their ascent in terraced ranks over sandy bays and rice fields; and a series of coastal mountain passes lead through spectacular highlands and coastal plains toward the ancient capital, Hue. Along the old riverfront tree-lined boulevard of **Bach Dang** are street-side cafés and marvelous old colonial villas from the time when Da Nang was a busy French trading port known as Tourane.

WHAT TO SEE AND DO

Almost every visitor to Da Nang stops at the **Cham Museum**. Located at the intersection of Tran Phu and Le Dinh Duong, it is a world-famous cultural attraction. This well-planned series of open pavilions houses the world's best and most diverse exhibition of Cham relics and sandstone carvings of Shiva, Brahma and Vishnu, alongside carved altars, lingas, Garudas and other Hindu symbols. The mood of most of these seventh- to fifteenth-century statues and sculptures is sensual, reflecting the essential role of fertility in this creed. Elsewhere, sculptures of apsaras, musicians and scenes from the Hindu epic *Ramayana* are reminiscent of the temple carvings of Bali and Java which share a similar heritage. The museum is open daily with a break from 11 AM to 1 PM for lunch.

Located in Hai Phong Street east of the railway station is **Chua Cao Dai** — the biggest and most important Cao Dai temple outside Tay Ninh Province in the south, serving the 20,000 or so Cao Dai followers who live in Da Nang. It was built in 1965 and, like the Great Temple of Tay Ninh, its colorful daily services are worth a visit.

There are three temples in the city area, the **Phap Lam Pagoda** in Ong Ich Khiem Street and the **Tam Bao** and **Pho Da** pagodas in Phan Chu Trinh Street. There's nothing very historical about them — the oldest was built in 1923 — but the Phap Lam Pagoda features a brass statue of Dia Tang, the God

ABOVE: Da Nang's splendid Cham Museum.
OPPOSITE: The Cham ruins in Singhapura.

of Hell, who most surely played a part in Da Nang's recent history, and is close to the Quan Chay vegetarian restaurant.

WHERE TO STAY

Da Nang has plenty of downtown hotels, but if you can get into one along the Han River embankment, you'll enjoy the view of one of Vietnam's busiest river ports. The old **Bach Dang Hotel** ((511) 823 649 FAX (511) 821 659 at 50 Bach Dang Street is one of the most comfortable, with air-conditioned double

rooms from US$28 to US$80 a night, including a buffet breakfast and a good verandah restaurant. At 142 Bach Dang Street is the **Tan Minh Hotel** ((511) 827 456 near the market. This popular riverside mini-hotel has only 10 rooms, some with a balcony, air-conditioning, and television, for around US$15 to US$30. Another riverside hotel at 36 Bach Dang is the **Song Hang Hotel** ((511) 822 540 with rooms from US$20 to US$ 80.

Out of town at China Beach is the new **Furama Beach Resort Da Nang** ((511) 847 888 FAX (511) 847 666, 68 Ho Xuan Huong, Bac My An, adding a little international *savoir faire* to the hotel scene, with rooms from US$98 to US$280. The restaurants are good too, especially the buffets.

WHERE TO EAT

A burgeoning nongovernmental organization population has helped Da Nang's restaurant scene considerably, resulting in several reasonable places to try. Popular with both expatriates and visitors, **Christie's Harbourside** ((511) 826 645 at 9 Bach Dang is owned by a New Zealander and serves a Western menu with pizzas, pastas, and seafood. **Café Indochina** ((511) 847 888 at Furama Resort offers around-the-globe cuisine and great buffets.

The **Bach Dang Hotel's restaurant** is quite pleasant — you can get a good Western breakfast with traditional coffee and a wide range of Western, Vietnamese and Chinese dishes at lunch and dinner. Also along Bach Dang Street at No. 7, is the **Hana Kim Dinh Restaurant** ((511) 830 024 with Vietnamese and oriental dishes.

The **Quan Chay Restaurant** at 484 Ong Ich Khiem Street, near the Phap Lam Pagoda, is a traditional Buddhist vegetarian establishment.

HOW TO GET THERE

Vietnam Airlines operates daily flights to Da Nang from Saigon. The Open Bus stops at Da Nang to visit the Cham Museum en route to Hoi An, and the city is on the main train route.

ENVIRONS

Marble Mountains

Eleven kilometers (nearly seven miles) south of Da Nang, the **Marble Mountains** Buddhist complex actually has five "mountains," all solid lumps of marble, each representing the five elements — fire, water, earth, wood and metal. The biggest, generally referred to as Marble Mountain itself, is pitted with grottoes containing stone carvings and more recently-erected stone Buddhas. During the Vietnam War, it was from here that Viet Cong soldiers would sit and watch the American GIs frolicking on the beach, a mere bazooka blast away. Getting up and down could be considered good exercise — it's a tremendous climb. Progress is somewhat hampered by the overenthusiastic child hawkers who relentlessly pester you every step of the way trying to get you to buy joss sticks and poorly-fashioned marble souvenirs. While Marble Mountain is hugely overrated, it is something of a must-see, and one of Vietnam's best-known attractions.

Take a taxi or try to pick up one of the truck taxis that await passengers outside the short-haul pickup truck station on Hung Vuong Street, a block from Con Market. If you care to brave the hassles of city bus

stations, buses frequently leave from the intercity bus station that go to Hoi An via Marble Mountains and China Beach. Tours are available from Hoi An.

My Son

The series of about 20 scarred Cham ruins at My Son, 60 km (37 miles) south of Da Nang along National Highway 1, was the key religious and intellectual center of Champa from the fourth to the thirteenth centuries. As such, it's probably the most important surviving relic of the Cham Kingdom, and one that prevailed over centuries of damage and pillage only to tragically become a battlefield during the Vietnam War. The Cham kings are thought to have been buried in a

complex which was probably as sacred to the Shivaite Chams as Angkor Wat was to the Khmer Hindus and Borobudur to the Javanese Buddhists.

The ruins lie in a valley near a coffee plantation and the towering **Cat's Tooth Mountain**. While nearly 70 former structures have been identified from the stone remains around the large site, only 20 give a clue today of what this complex once looked like. The main deity is Shiva, whom the Chams believed to be the creator and defender of the Cham Empire. The structures have been

categorized alphabetically into 10 groups by the archaeologists, but only Group B (tenth century), Group C (with an eighth-century shrine to Shiva, whose statue has been removed to the Cham Museum in Da Nang) and Group D (tenth century) have stood the test of time and conflict enough to satisfy the imagination. The site with the best preserved ruins was the eighth-century, now inaccessible Group A1, which was tragically annihilated by the American bombers. Group E (eighth to eleventh century), Group F (eighth century) and Group G (twelfth century) are part of a long-term restoration plan.

OPPOSITE: Worshippers at Marble Mountain, near Da Nang. ABOVE: Cao Dai prayer hall in Long Hoa.

The remaining structures include the eleventh-century stone base of a temple believed to have been first built seven centuries before; the walls of a tenth-century library with bas-relief brickwork of elephants and birds; a reasonably preserved tower; and a meditation hall earmarked for renovation as a new Cham museum. Restoration work has been going on among these sites, and the plan is to eventually restore other groups which have been reclaimed by the surrounding forest or were badly damaged during the war.

of Time. Close by, another far more modern church, the Mountain Church, overlooks the stone foundations of what was once Simhapura (or Singapura the Lion Citadel), the first Cham capital, from the fourth to the eighth centuries.

HOI AN (FAI FO)

Hoi An is one of those quirky backwaters that capture the imagination of all who visit. Situated 30 km (19 miles) from Da Nang via the Marble Mountains, this historic trading

My Son is another day trip by road from Da Nang, or better still, from Hoi An. If driving to My Son, the 35-km (20-mile) dirt road is about seven kilometers (four and a half miles) south of the Hoi An turn off located 27 km (17 miles) south of Da Nang. A reliable driver is essential, or else take a day tour from Da Nang or, preferably, from Hoi An. There is no accommodation in My Son.

Tra Kieu Museum

The 100-year-old Catholic church of this small town, some 20 km (12 miles) beyond My Son, is interesting enough — but it's also a museum of Cham relics collected from the local people by the priest. They include ceramic artifacts bearing the features of Kala, the God

port enjoyed for several centuries a prominence similar to that of Malacca and Macau as one of the great trading emporiums of the east. If you want to get a feel for the early days of Asian trading ports, this well-preserved ancient Cham city on the Thu Bon River is possibly the most authentic still existing in Asia. Hoi An has a long and patchy history that would make a marvelous background for a novel. Established early on, from the fourth to the tenth centuries Hoi An was the main trading port for the nearby Champa Kingdom (Hoi An is close to the citadels of both My Son and what was once Singhapura or Simhapura). Early ninth- and tenth-century Arab documents refer to the port, mentioning it as a provisioning stop.

After a long period of unrest between the Chams and ethnic Vietnamese, in the fourteenth century order was restored, and by the fifteenth century Hoi An or Fai Fo was attracting both Japanese and Chinese traders. As they stayed over in town, waiting for favorable monsoon winds, they settled in, establishing small communities and larger stocks of trade goods until, by the seventeenth century, a major trade transhipment center was established. When the Japanese left the town around 1637, by decree of the then Japanese emperor forbidding further contact with the rest of the world, the early Dutch, Spanish, Portuguese, English and French merchant ships started calling in, coming to trade for precious cargoes of quality silk, paper and textiles, molasses and areca nut, pepper and beeswax, porcelain and lacquer, mother of pearl and tea. Thai, Indonesian, Filipino, Arab and Indian traders all added both to the cultural melange and to the stock of items for trade.

While most of the old town was destroyed during the Tay Son Rebellion in the late eighteenth century, it was quickly rebuilt, and trade continued to flourish, right up until the end of the nineteenth century when, with the silting up of the Thu Bon River, trade gradu-

ally moved to the newer trading port of Da Nang (Tourane). By the early twentieth century, Hoi An had become a power of the past. It drifted into somnolence and nothing much changed until the early 1990s when it was discovered by backpackers and became very firmly placed on the tourist map. This has resulted in many of the town's well preserved homes, shop-houses, warehouses, and clan halls being restored and put to new use as enticing bars, restaurants, and galleries for the ever-growing stream of visitors. New mainstream hotels are springing up although

thankfully still on a small scale in keeping with Hoi An's architectural heritage.

WHAT TO SEE AND DO

With over 800 structures recognized as possessing historical significance, Hoi An is an architectural treasure house. Walk the narrow streets of the old town and see the well-preserved remains of the old town — almost a living museum of eighteenth-century houses. Although some have been converted to new uses, the exteriors are unchanged, except for some sensitive restorative work.

OPPOSITE: The Cham ruins in Singhapura.
ABOVE: An old wooden bridge crosses the Cai River at historic Hoi An.

The very active **Hoi An Tourism Company** ((51) 861 373 or 861 332 FAX (51) 861 636, 6 Tran Hung Dao, has produced well-illustrated and informative brochures about Hoi An's heritage and sells tickets that provide entrance to some of the best of the old houses.

The ancient **Japanese covered bridge** was built by the Japanese in 1593. It once connected the Chinese and Japanese communities before the Japanese left Hoi An in the mid-seventeenth century, after their emperor forbade further contact with the outside world.

The main thoroughfare of **Tran Phu Street** has several worthwhile houses to look at. The most well-known is at **No. 77,** a private house almost three centuries old. At 176 Tran Phu Street the Cantonese **Quang Dong Communal House** was founded in 1786, while the older **Chinese All Community House,** founded in 1773, was used by the five Chinese communities who traded in Hoi An — the Cantonese, Fukien, Hainanese and Teo Chew — before they assimilated into the local culture. Also in Tran Phu Street are the **Fukien Assembly Hall**, which later became the Temple dedicated to Thien Hau, the Protector of Sailors and Fishermen, and the **Quan Cong Temple**, founded in 1653.

A 20-minute bicycle ride past *padi* fields and a lagoon brings you to the **Cua Dai Beach** where deck chairs and vendors await new victims.

Another of Hoi An's biggest attractions is the relaxing, timeless ambience where sitting, talking, or shopping makes a very pleasant diversion, followed perhaps, with a visit to the Cua Dai Beach or a day's fishing boat trip to **Cham Island**, 21 km (15 miles) from town. The boat leaves from the town's main jetty on Bach Dang Street. Hoi An is known for its quality tailoring and it is a rare person who leaves without at least one new addition to their wardrobe.

WHERE TO STAY

From just one or two tiny guesthouses in the early 1990s, Hoi An's accommodation scene has blossomed to include dozens of alternatives. Many visitors prefer to stay in Hoi An rather than in Da Nang, and accommodation ranges from tiny rooms in traditional houses to glitzy new mini-hotels, backpacker lodging, and a perfectly acceptable new hotel across the river, a short five-minute walk from the old town and market.

The low-rise **Hoi An Hotel** ((51) 861 362 FAX (51) 861 636, 6 Tran Hung Dao, is the biggest establishment, with a large garden setting on the edge of town and rooms from US$17 to US$100. The **Vinh Hung Hotel** ((51) 861 621 FAX (51) 861 893, 143 Tran Phu Street, has rooms from US$20 to US$45. Located in the center of the Old Quarter in a historical house full of atmosphere, this hotel is often fully booked, so book ahead. The **Pho Hoi 2** ((51) 862 628, a private mini-hotel, is located just across the bridge over the river at Cam Nam. This new and comfortable hotel is run by friendly eager-to-please, staff. The **Pho Hoi 1** ((51) 861 633 FAX (51) 862 626 is at 7/2 Tran Phu Street, close to the market, and has rooms for US$8 to US$20. Cheap mini-hotels have sprung up all over town. The **Phu Thinh** ((51) 861 297 FAX (51) 861 757, 144 Tranh Phu Street, has rooms from US$12 to US$45.

WHERE TO EAT

The eloquently named **Café des Amis** ((51) 861 616 at 52 Bach Dang is a popular restaurant with lots of rumors about chefs and food quality. Try it and decide for yourself. The **Thanh Restaurant** ((51) 861 366 at 76 Bach Dang Street, in an old house by the riverside, serves seafood and Western dishes. The **Han Huyen Floating Restaurant** ((51) 861 462 on Bach Dang Street serves Vietnamese and Chinese food in a romantic setting on the river. The **Huong Hu (Yellow River) Restaurant** ((51) 861 053 at 38 Tran Phu is a popular spot overlooking the river.

HOW TO GET THERE

Being a small backwater off the main highway, Hoi An is a little difficult to access by public transportation. From the Da Nang train station or airport it will require a US$25 taxi ride. The most cost-effective way is to take the Open Bus from Hue (coming from the north) or Nha Trang (coming from the south) which will drop you in the center of the old town. The modern air-conditioned coach costs around US$8 for the day trip, with photo-stops at beauty spots and a lunch stop.

QUY NHON

A busy little timber port, Quy Nhon is not so important in its own right, but it is the nearest coastal access to the Cham ruins of Thap Doi, Bhan It (or Thap Bac), Duong Long, and the former Cham political capital of Cha Ban (Vijaya). It also provides access to the northern Central Highlands and to centers like Plei

12 km (seven and a half miles) north of town. One of Central Vietnam's largest Buddhist monasteries, the **Nguyen Thieu** is also situated in this holy spot.

About 26 km (16 miles) north is the **Canh Tien** (Brass Tower) and the walled ruins of **Cha Ban** — the only remains of the former Cham political capital that existed here from the tenth to fifteenth centuries. Close by is the still operating **Thar Thap Pagoda**, built in the mid-seventeenth century within the ruins of the old capital. Three other elaborately-decorated thirteenth-century towers,

Ku and Kontum, home to many minority villages and site of many fierce battles fought between the Americans and infiltrating North Vietnamese units during the war.

WHAT TO SEE AND DO

There are three main sites of **Cham ruins** to visit around Quy Nhon, remnants of a past empire. Close to the city on the edge of town are the two quite spectacular towers of **Thap Doi**. Rising to heights of over 18 m (60 ft), these newly-restored eleventh-century towers symbolize male and female powers. Standing on a hilltop between two branches of the Kon River are the four twelfth-century "silver towers" of **Ban It** or **Thap Bac**, about

the **Thap Duong Long** (Towers of Ivory) can be seen about eight kilometers (five miles) beyond Cha Ban. Some of these lintels show certain similarities with the Khmer style of the Bayon temple at Angkor and some extraordinary animal gods.

About 37 km (23 miles) to the south of Quy Nhon is the somewhat surreal landscape of **Song Cau** where breezy, stilted restaurants serve excellent fresh seafood, making it a popular lunch stop for tour buses and anyone else in the know. Beyond the restaurants are beaches and the fishing village of Song Cau, which is also a basketwork center.

One of several idyllic, untouched beaches between Da Nang and Hue.

WHERE TO STAY AND EAT

The beachfront **Quy Nhon Tourist Hotel** ((56) 822 401, 8–10 Nguyen Hue Street next to the Tourist Office, has rooms for US$18 to US$40. Close by is the **Phuong Mai Hotel** ((56) 822 921, 14 Nguyen Hue Street, with rooms for US$5 to US$10, while the evocatively named **Hai Ha Hotel** ((56) 821 295 has rooms for US$20 to US$30. On the beachfront, the **Binh Duong Hotel** ((56) 846 267 or (56) 846 355, has thirty rooms.

The **Tu Hai Restaurant** on the third floor of Lon Market on Phan Boi Chau Street has an English-language menu and serves a variety of Vietnamese and seafood dishes from 6:30 AM to 10 PM. The **Dong Phuong Restaurant** opens from 6 AM until 11 PM at 39-41 Mai Xuan Thuong Street. You might also try the **Ganh Rang Restaurant**, about three kilometers (nearly two miles) from the city center along Nguyen Hue Street, which is set on stilts in gardens right on the seafront and reportedly was frequented by the wife of Emperor Bao Dai.

HOW TO GET THERE

Lying 238 km (148 miles) north of Nha Trang, it's not the easiest place to get to: its airport at Phu Cat, another notorious wartime base, is 36 km (22 miles) north of the city. If you are taking the Open Bus north or south, it is a simple matter to get them to drop you by the town. Otherwise a tour or rental car is in order. Quy Nhon makes a good base to explore the surrounding sites.

Both Saigon Tourist and Especen Travel offer excellent tours of the Cham Ruins. See TAKING A TOUR, page 67.

NHA TRANG

Nha Trang has become the Surfer's Paradise of Vietnam — a major resort area with a laidback and relaxed atmosphere where you can feel that you are truly on vacation. While not quite as glitzy and big as that Australian resort, it grew dramatically from a faded French colonial resort to a GI R&R center, only to fade into gloom as a Russian retreat. The past five years have seen it spring from

a jaded old resort to the beginnings of glamour, with the rumbles of a famous future. With the front beach in the center of town, Nha Trang is being transformed with new big hotels and good restaurants overlooking the sparkling South China Sea.

The six-kilometer (nearly four-mile) main beach rivals Thailand's Pattaya, and although the girly bars are missing, plenty of pleasant entertainment is available. To the north are even more beautiful beaches just waiting to be explored, while in the vicinity are 71 islands, some with reportedly pristine diving and snorkeling locations beneath the waters. Nha Trang has a maximum potential that has only just begun to be realized.

Behind the beach is a cultural backdrop that features relics of the ancient Cham

kingdom of central Vietnam. The city and its gently curving beach lie 448 km (278 miles) from Saigon on a promontory that runs south of the Cai River estuary. A beachfront road, Tran Phu Boulevard, extends south from the river, becoming Tu Do Street as it approaches the peninsula. The business and administrative district, including the Nha Trang Hotel, are to the north of the airport, close to the estuary.

WHAT TO SEE AND DO

Five nearby islands provide fodder for numerous day trips. **Hon Tre (Bamboo Island)** is the closest and biggest, reached easily by boat, while the adjacent, **Hon Mun (Ebony Island)** is a popular snorkeling spot. **Hon Yen**, named after the swifts, or *salangane*, that build their edible nests in caves on the island, is 17 km (10.5 miles) out to sea and is one of several islands famous for their annual harvest of swift's nests, which are the essential ingredient of bird's nest soup. **Hon Tam** has a café and deck chairs while the closest, **Hon Mieu (Cat Island)**, has a lively fishing village and several pleasant seafood restaurants overlooking the beach. Although no one will tell you this because they would rather sell you a tour, you can get there by public ferry from the ferry point just beyond the Bao Dai Villas (from where all the island tours leave) and for which the locals will try to extort ridiculous prices from foreigners.

Morning at the fish market in Quy Nhon.

At the other end of the town over the Xom Bong Bridge to the north are the **Cham Towers of Po Nagar**. Standing like sentinels on a rocky hilltop overlooking the Cai River and northern entrance to Nha Trang are four of the original eight Cham towers of Po Nagar (The Lady of the City), among the finest examples of Cham architecture in central Vietnam. The towers were built between the seventh and twelfth centuries, although originally the site was a Hindu place of worship dating from the second century. Among these surviving monuments, the North Tower is

West of the city center at 23 Thang 10 Street, the **Long Son Pagoda** is a far more contemporary relic. Built in the late nineteenth century, the pagoda lies in the shadow of a towering white Buddha seated on a lotus. The temple's most distinctive feature is the mosaic dragons which adorn its entrance and roofs. It is home to a small community of monks. Located on Thai Nguyen Street (which is the eastern extension of 23 Thang 10 Street), the Gothic **Nha Trang Cathedral**, with its stained glass windows and towering spires, looks medieval but is even more

the oldest but best preserved, featuring stone wall carvings of Shiva dancing to the accompaniment of musicians, a stone gong and drum and, in the lofty main chamber, a 10-armed stone image of the goddess Uma. There's a small museum close to this site exhibiting other Cham relics. The three other towers are less illustrious, but two of them have stone lingas in their main chambers and the other, the Northeast Tower, features more bas-relief sculptures. The North Tower was built in 817 after the original temple complex had been destroyed by Indonesian raiders. As this book went to press, the towers were being restored in a project that is scheduled to last until mid-1999, if not until the end of that year.

contemporary — completed in 1933. Mass is held twice daily at 5 AM and 4:30 PM.

Nha Trang is also the proud home of a **Pasteur Institute**, one of three in Vietnam (the others are in Saigon and Da Lat), founded by Doctor Alexandre Yersin (1863–1943) in 1895. Born in Europe, Doctor Yersin came to Vietnam in 1889 after isolating the microbes that caused the bubonic plague in Hong Kong in 1894. After spending several years traveling and making observations in Vietnam, he introduced both quinine and rubber trees to the country. While the Institute is woefully short of funding, the dedicated workers carry on regardless.

The **Bao Dai Villas**, about six kilometers (four miles) south of town, are also worth a visit, or better yet, a stay. Bao Dai was the last emperor of the Nguyen Dynasty. He attempted to form an anti-Communist state still linked with France during the first Indochinese War, then abdicated when the Viet Minh achieved victory. His five seaside villas, built in the 1920s, are just north of the Oceanographic Institute and testify to the lifestyle which he obviously feared was threatened by the nationalist forces. President Nguyen Van Thieu took advantage of them when he was in power, and after the 1975 victory, they became resort rewards for the Communist hierarchy. They are now tourist guesthouses, renovated and air-conditioned, with lush gardens and excellent views of Nha Trang and the nearby islands.

Just below the villas, the **Oceanographic Institute** in Cau Da village was founded in 1923. Within the Institute are various tanks holding a wide variety of live marine life (including live seahorses), as well as major displays of 60,000 varieties of preserved marine life.

Recreational pursuits in Nha Trang include **diving** and **snorkeling** on the islands, and while recreational diving is still in its infancy in Vietnam, several companies are operating in Nha Trang. With 71 islands to choose from, this burgeoning industry has great potential (see SPORTING SPREE, page 32). Daily boat trips to the islands are also available. Ask at your hotel for the best operator or book a day trip from the Ana Mandara Resort for a more upmarket version. Getting around in Nah Trang is easiest if you rent a motorbike from your hotel, which should cost around US$5 per day. If a motorbike holds no appeal, there are always cyclos, which travel at a relaxed pace, or cars which also can be rented from the hotel.

WHERE TO STAY

Newest and nicest is the **Ana Mandara Resort** ((58) 829 829 FAX (58) 829 629, right on the beach at the south end of town on Tran Phu Boulevard. The chalet-style accommodation is furnished with natural materials, timbers and rattans and surrounded by gardens. Room rates range from US$137 to US$263. The towering **Nha Trang Lodge** ((58) 810 500 or (58) 810 700 FAX (58) 828 800 or (58) 829 922 is one of the first of a series of glittering highrise hotels planned for Nha Trang in the coming years. Looking across to the sea at 42 Tran Phu Street, the hotel offers plenty of facilities, balcony rooms and moderately expensive rates from US$75 to US$145 and higher. **Bao Dai Villas** ((58) 881 048, 049 FAX (58) 881 471, at Cau Da just to the south of town, enjoy the most dramatic setting in Nha Trang — a surviving colonial-style complex where you can get a feel for the grander days of Indochina. Huge, air-conditioned double rooms cost US$25 to US$80 a night, but don't expect facilities like a pool. You'll find the villas close to the tip of the peninsula, with views of the sea and the fishing fleet.

In the city center, the seven-story **Nha Trang Hotel** ((58) 822 347 FAX (58) 823 594 at 129 Thong Nhat Street has rooms from US$8 to US$25. While quite well run, it is a casualty of the free enterprise that is taking over the economy — government concerns just can't keep up.

The **Grand Hotel** ((58) 822 445 FAX (58) 825 395, 44 Tran Phu Street, is one of Nha Trang's gracious relics from a colonial past — what a shame the very communist staff haven't absorbed any of that graciousness. It badly needs a foreign joint-venture to restore its former grandeur. And the large rooms are overpriced. A cheap and cheerful hotel is the **Dong Phuong Hotel** ((58) 825 986 or (58) 828 247 FAX (58) 825 986, two streets from the beach on Hoang Hao Tham Street. A friendly, family atmosphere and big, clean air-conditioned rooms for US$20 to US$30. Fan-cooled rooms are much cheaper.

WHERE TO EAT

Along the beachfront are cafés and small restaurants specializing in seafood dishes, and the marvelous **Sailing Club** ((58) 826 528 at 72 Tran Phu, an entertainment/watersports/restaurant complex run by an Australian entrepreneur who saw a niche and jumped in. This popular upmarket venue

Fishing nets laid out for repairs at Hai Duang near Nha Trang.

serves Italian food with pizzas until 2 AM; has music in the big, open bar; serves quality mixed drinks; and has pleasant places to lounge. It is the place to go in Nha Trang.

The restaurant at **Bao Dai Villas (** (58) 881 049, just out of town, has superb sea views, great Vietnamese food at reasonable prices, pleasant service, and Vietnamese-style basic surroundings which can be ignored if you sit and gaze at the view. The chic **Ana Mandara Resort (** (58) 810 700 offers an excellent buffet, a dining room and poolside snacks. Close to the market is the **Dua Xanh** or **Coco Vert Restaurant (** (58) 823 687 at 23 Le Loi Street, which serves breakfast, lunch, and dinner, with live seafood and Vietnamese specialties. The **Casa Italia (** (58) 828 964, at Huong Duong Center in Tran Phu Street, offers pizza and pasta in a chic beachside setting. **Vinagen (** (58) 823 591, at the junction of Le Thanh Ton and Tran Phu Street, has Western and Vietnamese food and also sells the best tours in town. **Shanti (** (58) 825 944, inside the Vinagen compound, serves authentic Indian food by the beach. A popular outdoor pavement eatery in the evening serving fresh seafood is the **Truc Linh Restaurant (** (58) 825 742, open from 6 AM until 11 PM at 12 Biet Thu Street.

HOW TO GET THERE

While most visitors regard Nha Trang as a stop off point on the north-south route, it is quite feasible to spend a whole vacation here at the beach. Vietnam Airlines operates regular flights to Nha Trang from Saigon, Da Nang, and Hanoi but be aware they are subject to change without notice if a higher priority request comes through. The Reunification Express stops at Nha Trang and it is also a stop off for the north south Open Bus route.

ENVIRONS

One of Vietnam's more beautiful beaches is **Doc Let Beach**, 40 km (25 miles) north of Nha Trang past Binh Hoa. This long stretch of white sand flanked by casurinas is idyllic and untouched for most of the year (don't go on a Vietnamese public holiday) and most days you can have the beach to yourself. Simple

accommodation is available along with a restaurant with changing rooms. To get there, rent a motorbike or car from Nha Trang or take a *moto* from the turnoff at Binh Hoa. Another pleasant possibility is to rent a car or motorbike and head north to explore other beautiful beaches that line this magnificent coastline, or take a trip out to Monkey Island. The ferry point is about 20 km (13 miles) north of the city.

DA LAT

A French town transplanted into the foothills of Vietnam's Central Highlands, the "City of Eternal Spring" has a faded charm and crumbling French architecture that have endured through all the hard years of Communist rule. The bad old days are long since gone and today's business is tourism. Nestling in a valley 1,475 m (4,800 ft) above sea level in the south-central highlands, it makes a cool and otherworldly respite from the rigors of Vietnam. Developed by the French at the suggestion of Alexandre Yersin as a cool, high-country retreat from the obsessive summer heat of Saigon and the Mekong Delta, the hilly town is filled with elaborate villas, most of which attempt to recapture the familiar architecture of Normandy and Brittany. The French also turned part of the city, now known as the French District, into a complete replica of a provincial home town, and strolling the more beautiful parts in the cool evening, it is easy to consider that they didn't do a bad job.

Much of the city surrounds the hugely ugly Central Market, with hotels and the surviving French villas set on ridges and hillocks around it. Da Lat's centerpiece is the picturesque and newly-renovated Xuan Huong Lake — a reservoir lake created by the French in 1919. To the north of the lake, a sweeping low-rise hill leads to Da Lat's mecca for honeymooning Vietnamese, the Valley of Love. Surrounding Da Lat are acres of market gardens supplying vegetables and flowers to both Da Lat's and Saigon's markets. One of Da Lat's specialty crops is strawberries which can be bought dried, as delicious jam, and even fresh.

Da Lat has one of Vietnam's most glamorous hotels and while a sojourn there comes at a premium price, its chic and quiet grandeur make the Sofitel Da Lat Palace a special place. A recent renovation has revitalized this classic 1920s French hotel, restoring it to its original grandeur.

Small pony traps ply Da Lat's downtown streets, and are an ideal way to get around the inner city. But once the poor thin-shanked beasts start to climb the hills, the instinct is to jump out and give them a break. Bicycles are another ideal form of transport, but again, once you hit the steep hillsides you'll need to jump off and give yourself a break. Motorbike taxis are available at every corner and ladies can ride sidesaddle. Check where to rent bikes at your hotel desk. They will either have them available or know a friend who has. The major hotels provide cars and minibuses, and this is the comfortable and convenient way to go any distance, especially to the viewpoints, waterfalls and picnic spots in the hills around the city. Otherwise, Da Lat is a city made for strolling, and compact enough not to walk you off your feet.

WHAT TO SEE AND DO

Da Lat is a Vietnamese honeymooner's paradise, hence the romantic names — Valley of Love, City of Eternal Spring, the Lake of Sighs. Da Lat is also the center of a booming sex industry.

The **Valley of Love** is a major Vietnamese tourist attraction so think kitsch — tacky doesn't quite describe it. This very commercialized souvenir complex, also known locally as the "Valley of Shopping," is popular amongst Vietnamese newlyweds, where stalls are packed with stuffed wildlife of every description from the forests around Da Lat. It's an environmental nightmare, and flies in the face of every effort that Vietnam is now making to preserve the wildlife that survived years of bombing and chemical defoliation during the war. The Valley of Love is also the habitat of Da Lat's tourist touts — young men dressed as cowboys offer kids pony rides around the hills.

On the other hand, the **Da Lat Golf Course** has to be one of the more beautiful courses found in Asia. Misty mornings render it almost invisible, then as the sun melts it away,

colonial edifices, such as the spire of the cathedral, appear across the rolling hills. The 18-hole course is open to guests of the Da Lat Palace Hotel and arrangements can be made for other guests.

The inveterate temple-goer will be relieved to know that in addition to the pink **Da Lat Cathedral** on Tran Phu Street, several Chinese and Vietnamese temples can be seen around the city. One of the most popular temples is the **Thieng Vuong Pagoda** on Khe Sanh Street about five kilometers (three miles) from the city center.

Almost every visitor to Da Lat makes it to the **Central Market** at least once during their stay. Within the massive concrete neo-revolutionary construction are magnificent fresh flowers and vegetables, strawberries of all descriptions, including the delectable Da Lat strawberry jam, clothing, and market things. On the top floor is a food center with numerous stalls selling Vietnamese specialties, juices, and some vegetarian food. Around the market are stalls selling more fruits and flowers while at the top of the steps are a few women selling tribal artifacts of a poor quality, but it's always fun to look.

Da Lat University, 1 Phu Dong Thien Vuong Street at the corner of Dinh Tien Hoang Street, and the **Domaine de Marie Couvent**, commanding a fine view of the town from its hilltop perch at 6 Ngo Quyen Street, are other landmarks, as is **Bao Dai's Summer Palace**, on the outskirts of town in

Vendor prepares fruit display in Da Lat's central market.

a grove of pine trees off Le Hong Phong Street, a villa constructed in 1933, it is not one of his most beautiful hideaways. If you can run the gauntlet through hordes of souvenir stalls, armies of Disney characters, "cowboys" on ponies and troops of photographers to get into the place, for a fee you can explore the villa whose 25 rooms are full of artifacts. Not everyone's cup of tea but a very popular attraction, the Gaudiesque cement **Crazy House** or tree houses attracts hundreds of visitors a day. Designed by Soviet-trained architect Doctor Dang Viet

Nga, the house is really a hotel with tiny themed rooms where you can spend the night. While her creation is no doubt constructionally clever, it is the tackiness that one's breath away — this is kitsch of the highest degree.

WHERE TO STAY

The five-star **Sofitel Da Lat Palace Hotel**, ((63) 825 444 FAX (63) 825 666, at 2 Tran Phu, is really the only place to stay for those on an unlimited budget. Beautifully renovated in 1993 by a joint-venture company, and under the management of Sofitel, the 78-year-old property isn't a big hotel — it only has 43 charming rooms and suites. The facelift

remained faithful to its old French style and fittings, and it has the best restaurants and bars in Da Lat. In a parallel renovation, the same company has upgraded some 17 French villas in Da Lat as tourist chalets and has built Vietnam's first international-class 18-hole golf course north of Xuan Huong Lake.

Another hotel with a long tradition is the 143-room **Novotel Da Lat Hotel** ((63) 825 777 FAX (63) 825 888 at 7 Tran Phu Street, opposite the Palace, which opened its newly renovated doors in December 1997. The hotel

dates back to 1907 and while very pleasant, it lacks the charm and enormous rooms of the Palace. Rooms cost from US$119 to US$189 for suites.

The **Minh Tam Hotel** ((63) 822 447, at 20A Khe Sanh Street, about three kilometers (just under two miles) from the city center, is distinctive for its peaceful location amid wooded hills and its history — it was the summer villa of the infamous sister-in-law of the President Ngo Dinh Diem, Madame Nhu. Built in 1936 and renovated in 1984, rooms cost US$25 to US$35. The **Anh Dao** ((63) 822 384, 52 Hoa Binh Street, has rooms for US$30 to US$50.

Da Lat is said to have about 2,500 old villas, but not many of them are in any condition to serve the tourist market, and the company that was operating many of them has ceased. For current status ask at the **Lam Dong Tourism Company** ((63) 822 125 or (63) 821 351 FAX (63) 828 330 or (63) 822 661, at 4 Tran Quoc Toan.

WHERE TO EAT

The Sofitel Da Lat Palace offers the most sophisticated dining room in town while the animated **Shanghai Restaurant** ((63) 822 509 at 8 Khu Hoa Binh Street, close to the Central Market, has good Vietnamese, Chinese and Western food. A popular lakeside restaurants is the **Thanh Thanh** ((63) 821 836, at 4 Tang Bat Ho, for traditional Vietnamese cuisine and Western dishes popular with tour groups. A family restaurant, the **Maison Long Hoa** ((63) 822 934 offers a cozy atmosphere and good Vietnamese and Western food at 6 Duong 3 Thang 2 (Duy Tan), Da Lat. On top of the market is the easily visible and not terribly beautiful **Da Lat Club** which looks out across town. It is good for a drink around sunset, sitting on the terrace watching the lights of town come on.

ENVIRONS

On the outskirts of town on the Nha Trang road, the **Chicken Village** is marked by a huge chicken statue at its entrance. A rudimentary K'Ho village, it's known for its quality weaving. Lining the entrance to the village are stalls selling Cham and K'Ho handlooms, many in silk. While they are brought in from other weaving centers and made for the tourist market, the quality is quite good, and there are some nice pieces available for around US$10.

While Da Lat is blessed with numerous, once-beautiful waterfalls, they too have become major attractions, clothed in concrete, kiosks and more ponies and Disney characters — explore at your own risk. One of the nicer waterfalls is the **Datania Falls**, about 20 km (13 miles) from town along Highway 20.

Out of town are coffee plantations, highland walks and minority villages to visit. **Lang Bian Mountain**, at 2,162 m (7,095 ft),

makes a good day's hike. Leaving from the village of **Lat** about seven kilometers (four and a half miles) north of Da Lat, the walk takes three to four hours, passing through many minority villages of four different ethnic groups along the way. Young local boys are happy to act as guides for a reasonable fee.

HOW TO GET THERE

Daily Vietnam Airlines flights operate to Da Lat from Saigon (the airport is 30 km or

19 miles south of the city) but the best way of arriving is by road, either up the incredibly panoramic road from Nha Trang, or the scenic, inland eight-hour drive from Saigon. After crossing wide agricultural plains, the elevation rises, passing through tea and coffee plantations before reaching the densely forested hills that surround Da Lat. Once there, you'll find that the city meets every description that you may have read elsewhere — it certainly enjoys a sense of mountain magic, especially in the chill of the night when the heavens are brilliant with stars.

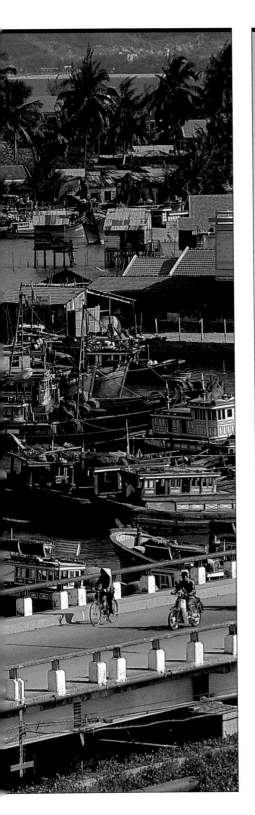

Southern
Vietnam

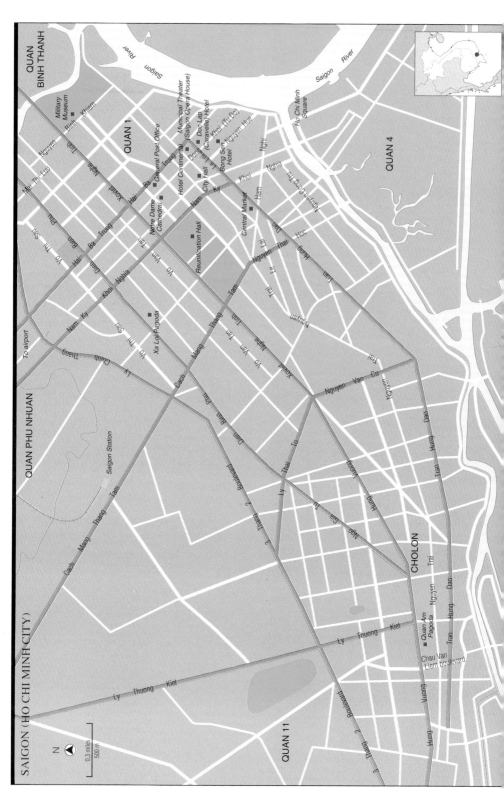

SAIGON (HO CHI MINH CITY)

N

0.3 miles
500 m

Saigon River

Saigon River

QUAN BINH THANH

QUAN 1

QUAN 4

QUAN PHU NHUAN

QUAN 11

CHOLON

Ho Chi Minh Square

Military Museum

General Post Office

Municipal Theater (Saigon Opera House)

Doc Lap

Khoi (Tu Do)

Hotel Continental

City Hall

Notre Dame Cathedral

Bong Sen Hotel

Nguyen Hue

(Caravelle) Hotel

Reunification Hall

Central Market

Xa Loi Pagoda

Saigon Station

To airport

Quan Am Pagoda

Chau Van Liem boulevard

Nguyen

Binh Khiem

Nguyen Thi Minh

Han Thuyen

Nguyen Du

Ba Trung

Han

Duy

Vo Thi Sau

Nam Ky

Khoi Nghia

Van Tran

Le

Ky

Nghia

Nam

Don Dat

Ly

Ham Nghi

Ham

Hoc

Lai

Nguyen Thai

Le

Loi

Trai

Nghe

Nguyen

Trai

Vo Van Tan

Dinh Tien

Nguyen Van Cu

Cong Quynh

Cach Mang Thang Tan

Chinh Thang

Vo Thi Sau

Ly Thuong Kiet

Cach Mang Thang

Dien Bien Phu

3 Thang 2

3 Thang 2

Nguyen

Ly Thai To

Ly

Thai

Hung

Ton

Dang

Tran Hung Dao

Tran Hung Dao

Ly Thuong Kiet

Nguyen

Trai

Hung Vuong

3 Thang 2

Boulevard

SAIGON (HO CHI MINH CITY)

What a contrast to the north! While Hanoi is sedate and orderly, Saigon is brash, entrepreneurial and fast — even the pickpockets are talented. While Ho Chi Minh City remains the official name for Saigon, southerners continue to call the city by the name they always have used: Saigon.

In the war years, fattened with billions of American dollars, the city was regarded as one of the richest, most advanced, and cer-

Equatorial, and Hilton are all making their mark in competition with the government-run French colonial hotels and well-loved names like the Rex, the Continental, the Majestic, and the newly renovated and opened Caravelle.

Each Saturday and Sunday evening, the broad square between the Hotel Continental and Municipal Theater becomes the center of an impromptu show as thousands of adults and teenagers ride their motorbikes down **Dong Khoi Street** to the riverfront. It's quite a spectacle, a continuous stream of

tainly one of the most sophisticated in Asia. It came under heavy suppression after the fall of 1975, many of its intellectuals and entrepreneurs — the ones that didn't make it to freedom in the United States, that is — were jailed or put into reeducation camps, and a system of high-security paranoia was imposed on the people without any significant long-lasting effect. Now, Saigon's former capitalists are back in favor, and this vital city has taken off into the stratospheres of a free-wheeling economy.

Saigon is the center where new hotels are at their glitziest and most luxurious, products of numerous joint-venture deals with international hotel groups from around the world. Omni, Hyatt, Ramada, New World,

headlights and roaring engines that flows until after midnight; and there's an interesting story to it. According to some reports, this is a social outing that began as a weekly demonstration, a show of force, by the Saigonese to remind the city's Communist authorities and the government in Hanoi that this is a city not to be messed with.

Another pleasant pastime which gives a feeling of old Saigon is to sit back in a comfortable rattan chair and enjoy a late afternoon beer or *citron pressé* on the roof of the Rex Hotel, one of the must-visits on any Saigon itinerary. Stay on for dinner and enjoy one of the most pleasant spots in town.

ABOVE: Saigon's Opera House flanked by the famed Hotel Continental in downtown Saigon.

Breakfast amongst the chirping caged birds on fresh and crisp baguettes, *café filtré* and eggs is a memorable experience.

Saigon runs on foreign investment, its downtown streets and markets packed with imported luxuries and the latest Japanese, Korean, and Taiwanese consumer appliances. The traffic is a heady mix of motorbikes and cars, taking over the bicycles and pedal-powered cyclos of yesterday, and crossing the road can be a big adventure. Traffic lights don't play much part in Vietnamese traffic. Everything *merges*, a weaving of vehicles at major

intersections. Pedestrians too, must learn how to merge, which means taking a deep breath and stepping into the street looking straight ahead — any sign of weakness and you'll be swamped by the traffic. Motorists can pick a beginner a mile off, and for starters it is best to follow a Vietnamese veteran until you get the hang of it when, magically, the traffic just melts around you, giving an enormous sense of achievement.

Unlike her northern sister city, Saigon is an entertainment hub with bars, discos, clubs and karaoke, and other less salubrious establishments operating far into the night. Although the sprawling suburbs are booming with new housing and joint-venture factories, the inner city is the main tourist area

which sprawls between District One — the downtown business area which extends southeast from the elegant Notre Dame Cathedral to the Saigon River, centered on Dong Khoi Street (known as Tu Do Street during the war) — and District Five, the teeming Chinatown, to the west.

GETTING AROUND

Dong Khoi Street, Nguyen Hue, and Hai Ba Trung are where most of the key commercial centers, travel agencies, airline offices, shops, banks, cultural centers, and hotels are located. The Central Market (Ben Thanh), lying west of Dong Khoi on Le Loi Boulevard, and the imposing New World Hotel front a wide traffic circus from which the main access streets to Cholon radiate.

When I say that Saigon is still quaintly antique, I mean that the downtown area most visitors today will be familiar with is very much what it was during the war — surprisingly untouched by the spate of development that has enveloped the city. An everyday itinerary takes in Notre Dame Cathedral and the adjacent Post Office, an architectural masterpiece in its own right, and extends south down Dong Khoi to the stately Hotel Continental and the newly-renovated Municipal Theater, home to the popular Q Bar, west past the Eden Center and public square to the Rex Hotel and the Hôtel de Ville (City Hall), then on to the Central Market, south again via the Doc Lap (Caravelle) Hotel on Dong Khoi, or a series of mid-street photo and souvenir kiosks on Nguyen Hue to the Saigon River waterfront.

While this is not all there is to the city, it is certainly the hub of most activity with virtually everything that any visitor would need. Although the hotels and travel agencies in Saigon have limousines, cars and minivans available, the pedal cyclo is really the only way to get around the city. When I was there, they were charging the equivalent of US$1 for a short trip, but I would advise you strike a deal and hire one for a few hours at US$1 per hour. It's best to hire one around

OPPOSITE: Midday traffic rolls by the landmark Notre Dame Cathedral, Saigon. ABOVE: The graceful lines of City Hall and Ho Chi Minh Square in central Saigon.

your hotel, and let the concierge know which one it is. Like Hanoi, most of the drivers are war veterans, and most are helpful and polite — but you occasionally run into one who'll hassle you for extra money while you're on the road.

Motorbike taxis are available for quick visits to the bank and other tiresome necessities, or rental car drivers will generally materialize outside your hotel. **Hire A Car — Mr. Trung** pager number (08) 130 8446 or mobile (090 709 732; can find you a four-wheel drive vehicle; four-, seven- and fifteen-seater cars; and Lexus, Hondas, or Toyotas. "We are always at your service with reasonable price" is his motto. **Motorbikes For Rent** can be found at De Tham and Bui Vien Streets, District 1, with prices for a 50cc Honda Dream starting at US$4.50 a day and working up to US$10 a day for bikes over 70cc.

GENERAL INFORMATION

Saigon tourism has generally been operated by government monopolies, until quite recently when quite a few private companies have begun operations. This has been beneficial for consumers who are no longer forced to take overpriced and fairly stuffy tours. That said, **Saigon Tourist** ((8) 829 8129 FAX (8) 822 4987, 49 Le Thanh Ton, District 1, have an excellent range of tours and can be quite helpful with information over the counter. Other tour companies are listed under TAKING A TOUR, page 67.

WHAT TO SEE AND DO

Central Saigon
Notre Dame Cathedral makes a useful starting point for a colonial city tour. This elegant blend of red brick and white stone with two soaring, spired towers faces directly down Dong Khoi toward the Saigon River. Built in 1877, Notre Dame and its spires, viewed across a green cloud-bank of trees, add an almost pastoral touch to Saigon's fast-rising skyline. The cathedral fronts a sweeping square which features a tall white statue of the Virgin Mary, and a well-kept tree-shaded public park sprawls to its rear. Closed by the Communists after the 1975

takeover of Saigon, Notre Dame is now flourishing again, with services beginning at dawn each day.

Adjacent to Notre Dame, the **General Post Office (GPO)** is another striking architectural landmark, built in 1886. It combines an opulence of colonial bas-relief and charming shuttered windows with a vast central hall reminiscent of a canopied Victorian railway station. Today, it's still the main communications center for the Saigonese, but the hotels have now usurped it as a provider of telephone, fax and postal services for visitors. The GPO is also home to a pack of virulent and talented pickpockets, and while there are periodic sweepings by the police, they quickly return so visitors are warned to beware.

Close to Dong Khoi Street, in fact, right on the corner of Dong Khoi and Le Loi Boulevard, the **Givral Patisserie and Café**, with its big picture window's view of the Opera House, is an old Saigon institution and still worth a visit for a *café filtré* or a *citron pressé* and the odd meal. Veterans will remember it for its afternoon coffee and ice cream, while newer visitors will enjoy the same service, no doubt from the same old veteran waiters. The stately old *grande dame* of Saigon, the **Continental Hotel**, just adjacent to Givral Café, renowned for evening drinks and the shoeshine boys in its open verandah bar during the war, is as much a cultural attraction as a hotel. Somehow, it's managed to tart itself up yet come down a peg or two — the famous verandah bar is

now closed off and turned into an airline office, and the standard of service has lowered since it became a state-run establishment. Still, it's something of a treat to breakfast on hot croissants, baguettes and coffee in the inner garden courtyard, once favored by foreign correspondents. No doubt if an international hotel management group takes over, it will be restored to its former splendor.

Right across from the Continental and facing Le Loi Boulevard, the **Municipal Theatre**, or **Saigon Opera House**, as it's more commonly known, with its high, arched entrance, was South Vietnam's National Assembly during the war. Now it's the city's

Saigon's Municipal Theatre was the often stormy legislature of South Vietnam during the war.

main cultural center, with concerts, recitals, plays and performances by Saigon's new crop of pop artists frequently staged there. During the war, the gardened square opposite this building featured a huge, grotesque concrete sculpture depicting a United States soldier and his ARVN (Army of the Republic of Vietnam) comrade charging into battle. But cynicism was so rife among correspondents and United States troops that it was regarded as a United States soldier pushing his Vietnamese counterpart into the fight. Even more cynical was another popular

mess with a stage for regular go-go dance shows, and a viewpoint from which visitors could watch the war at night — the tracers from jet cannon and helicopter machine-guns pouring into the city's outer suburbs.

Today, the Rex's rooftop is being dwarfed by a new rash of high rise buildings, but it remains immensely popular. One of the can't-be-missed exhibits is the bizarre exhibition of fiberglass Disney-style elephants, deer, and other decorations interspersed with trees cut to the shape of animals. Its open bar is still a good position from which to view the city's

unofficial description — a United States soldier and ARVN comrade attacking the National Assembly.

The charming, renovated **Rex Hotel** on Le Loi Boulevard, facing Ho Chi Minh Square, also deserves special mention for its wartime role. This was the Rex BOQ (Bachelor Officer's Quarters for United States personnel) during the war, heavily sandbagged against Viet Cong attack and one of the focal points of the United States and international media. Its ground floor, occupied by JUSPAO (Joint United States Public Affairs Office), was where the famous "Five O'clock Follies" — a daily media briefing on the war — was staged. The establishment's rooftop was a sprawling officer's bar and

fast-changing skyline. The Rex is the one hotel in Saigon that I feel gives a real feel of Vietnamese hospitality, and as one of the managers confided to me, it is Saigon's most profitable hostelry with no need for an advertising budget or marketing ploys, the hotel sells itself.

Facing Ho Chi Minh Square and a bronze statue of "Uncle Ho" embracing a child, the excellently-restored colonial building of **City Hall** is one of the city's prettiest. Once the Hôtel de Ville, it is now the headquarters of the People's Committee. Its opulently-decorated façade and clock tower provide a backdrop to the square, which is crowded weekends with Vietnamese sightseers and families taking snapshots. Those who know this location from the old days will be interested to learn that the Rex Cinema is still there, and is still virtually packed out most evenings, although it is soon to be demolished along with the garage to make way for yet another extension to the ever burgeoning Rex Hotel.

The huge hangar-like covered market complex of **Central Market (Ben Thanh)**, west on Le Loi Boulevard from the Rex Hotel, and close to the New World Hotel, has to be seen to be believed — words like "crowded" and "bustling" just do not do justice to this incredible bazaar. Shoppers come here in

Ham Nghi Market has become one of Saigon's main venues for consumer electronics, its shops and sidewalks piled with Japanese and Taiwanese televisions, video cassette recorders, hi-fi systems and computer equipment. In the streets between Ham Nghi and Nguyen Hue, a sprawling open market is also full of imported watches, video games and other electronic products. The main tourist run of **Dong Khoi Street** is another must. Along this tree-shaded thoroughfare are dozens of hotels, gift shops, pickpockets, and art galleries, as well as bars and restaurants.

their hordes to pick and jostle their way through stalls packed cheek-to-jowl and piled high with consumer and household goods, shoes, cheap clothing, poultry, fish, and just about every other product and provision you can name. The market has entrances on all four sides, so you and the pickpockets can get in quite easily. Getting out is another thing altogether, such is the general crush of shoppers, wandering hawkers, beggars, monks, and novices cadging alms, and the ever-persistent kids selling lottery tickets, maps, postcards, or books. Another teeming open market fills the streets around Ben Thanh, and around that there are shops and boutiques selling home electronics, watches, jewelry, luggage and fashion clothing.

Dong Du Street, which runs west from Dong Khoi from just outside the Bong Sen Hotel, is notable for three attractions — the **Saigon Business Center**, set up to provide modern office facilities and communications for incoming joint-venture companies; the **Central Saigon Mosque**, which thrusts its white minarets up through the general concrete decor of this district, and a series of new tourist bars centered on the **Apocalypse Now**. As the name suggests, Apocalypse Now is a long-running popular bar and a magnet for all those who knew Vietnam during the war. It also attracts the trendiest

ABOVE and OPPOSITE: Saigon street life — as in most cities in Indochina, two wheels get the job done.

of backpackers and local expatriates, a good place to blow your mind for an evening, cross-check travel information, and contemplate the adventures you will probably encounter on the way back to your hotel — grappling with pickpockets, motorcycle thieves, and the large crowds of beggars and homeless children, hawkers, and cyclo drivers who tend to rush you the moment you step out to go home.

All main streets in this central downtown area eventually converge upon the vast piazza of **Me Linh (Hero) Square** next

Museums

Another must on a Saigon tour is actually called the **War Crimes Museum**. This complex of old buildings at the intersection of Le Qui Don and Vo Van Tan Streets is an emotionally debilitating exhibition of American and allied atrocities committed during the Vietnam War. The My Lai massacre, the Phoenix Program torture sessions, the Viet Cong suspects being hurled from helicopters — it's all there, presented in room after room of photographic displays. It says nothing, obviously, about Commu-

to the Saigon River, dominated by a towering statue of the hero himself, Tran Hung Dao, who led the resistance against the invading Mongols in 1287. The square leads to the **Saigon River** with a gaggle of tourist cruise boats and the elegant **Majestic Hotel**.

To the west, the waterfront is dominated by the city's naval headquarters and a huge dry-dock which spends most of its time repairing battered Russian freighters. Hero Square is where you can rent a launch to tour the **Port of Saigon**, cruising among the container ships and something that's now quite unique to southern Vietnam — high-prowed lighters and cargo junks with big eyes painted on their bows.

nist atrocities, and it seems a shame that this litany of crime and suffering could not have embraced the mutual savagery and collapse of human values that distinguished the Vietnam War from other contemporary conflicts. What particularly struck me amid this harrowing scene was the frame and rotors of a helicopter gunship nestled under a tree in front of the museum, a huge Long Tom artillery piece — which once hurled shells across the Demilitarized Zone (DMZ) into North Vietnam — nearby, and, alongside that, one of the guillotines that the French imported to deal with nationalist activists. Other than that, it was fascinating to view news pictures of the political giants of the Vietnam War — Lyndon B. Johnson,

Dean Rusk, Robert McNamara, Ambassador Elsworth Bunker, General William Westmoreland, Presidents Nguyen Cao Ky and Nguyen Van Thieu — and to contemplate how ignominiously they've passed into history since.

Located in what was once Gia Long Palace, the **Museum of the Revolution,** built in 1886, is a typical Indochinese shrine to the Communist triumph, with all the politically correct, pro-Soviet paraphernalia that you'll find in similar museums in Vientiane and Phnom Penh. Also located

Thieu was finally forced from office shortly before the fall of 1975. It's also significant as a symbol of the Communist triumph — you may recall the television news scenes of the tank crashing through the palace's main gates, and the Viet Cong soldier rushing to fly the revolutionary flag from its balcony. Nowadays, Reunification Hall is a tourist attraction, open all days except Sunday, and a trade exhibition center. Ex-President Nguyen Cao Ky, once famous for his dashing flight-suits, pearl-handled revolvers and equally flamboyant First Lady,

on Nguyen Binh Khiem Street, at the zoo entrance, the **History Museum** traces Vietnamese history from the Bronze Age to the post-1975 era, and includes a research library. Prior to the fall of Saigon it was the National Museum of South Vietnam, built in 1929 by the Société des Études Indochinoises.

The contemporary building of **Reunification Hall**, erected in 1966, is a blend of Western and Asian design, lying in sweeping grounds surrounded by a decorative steel fence beyond the public park to the west of Notre Dame Cathedral. It's a significant landmark — it was the Presidential Palace, and home of Nguyen Van Thieu, from the height of the Vietnam War until

ended up running a store in Los Angeles. As for Thieu, he died a virtually forgotten recluse in England.

More than 1,000 varieties of orchid can be enjoyed at the **Orchid Farm,** including two named after Richard Nixon and Joseph Stalin, located 15 km (nine miles) from downtown Saigon along the highway to the former United States military base at Bien Hoa. Visits can be arranged through your hotel desk or travel agency, and the best time to go there is in January or February when the plants are in full bloom.

OPPOSITE: Downed United States jet in the Revolutionary Museum, Saigon. ABOVE: Saigon's Ho Chi Minh Memorial rests alongside the Saigon River port.

Temples

A great many Buddhist and Taoist temples can be seen in Saigon and Cholon, some of them neglected and dilapidated, others gleaming with garish renovations. The oldest is the **Giac Lam Pagoda**, northwest of Cholon at 118 Lac Long Quan Street, which was built in 1744 but completely restored in 1900. This temple and the nearby **Giac Vien Pagoda** at 247 Lac Long Quan Street, dating back to the late eighteenth century, both feature a great many relics and images including white statues of the Goddess of Mercy.

Xa Loi Pagoda, at 89 Ba Huyen Thanh Quan Street, north of Ben Thanh Market, is probably most significant because it was here that dissident monks immolated themselves in protest against the Diem regime in the early 1960s. The ritual suicides followed a raid by government forces in 1963 in which the temple's 400 monks and nuns were arrested. Another quite historical place is the **Tran Hung Dao Temple** at 36 Vo Thi Sau Street, dedicated to the hero of Vietnam's battle to defeat the invading Mongols of China's Yuan dynasty.

Two much newer temples are the **Dai Giac Pagoda** at 112 Nguyen Van Troi Street on the way to the airport, and the **Vinh Nghiem Pagoda**, built in 1971 with the support of the Japan-Vietnam Friendship Society, whose eight-story pagoda towers over the suburban sprawl of Nam Ky Khoi Nghia Street. Among the Chinese temples of Saigon and Cholon, the **Emperor of Jade Pagoda** at 73 Mai Thi Luu Street is the biggest and most opulent. Dedicated to both the Buddhist and Taoist creeds, it was built in 1909 by Saigon's Cantonese community, and its courtyards and three main prayer halls are crowded with images of the divinities, including the Goddess of Mercy, the Sakyamuni Buddha, and the Emperor of Jade himself. The city's Fujianese immigrants built their main shrine, **Phung Son Tu Pagoda**, at 338 Nguyen Cong Tru Street, and it's dedicated not only to the Goddess of Mercy but the Guardian Spirit of Happiness and Virtue as well.

In Cholon, the renowned **Quan Am**, at 12 Laos Tu Street, is generally recognized as the district's most important shrine, but you'll find a far more picturesque temple, the **Thien Hau Pagoda**, dedicated to the Goddess of the Sea, in crowded Nguyen Trai Street. Above the entrance to this small but colorfully restored temple there's an elaborate ceramic frieze featuring mandarins, immortals and scenes from the Taoist legends. The main hall is reminiscent of the famous Man Mo Temple in Hong Kong — huge coils of smoking incense hanging from its rafters. Not far from this temple on Nguyen Trai Street, another Chinese place of worship, the **Nghia An Hoi Quan Pagoda**, or **Ba Pagoda**, features gilded bas-relief wood carvings and images of Ba Thien Hau and the Guardian Spirit of Happiness and Virtue. It was founded in 1760 by Chinese immigrants in thanks for guiding them to Vietnam.

Cholon District

During the Communist Tet Offensive of 1968, the Viet Cong infiltrated Saigon through the urban beehive of Cholon, populated by ethnic Chinese, to the west of the downtown area. And when you see its densely crowded streets you'll appreciate why. Cholon isn't just a hum of activity, it's more like a full-scale orchestra, and you'll find you're exhausted after a couple of hours battling its traffic and pedestrians. Chinese immigrants

ABOVE: Entrance to Thien Hau Temple in Cholon.
OPPOSITE: Street vendors outside Saigon's
Mariammam Hindu Temple.

began setting up shop here in 1778 and, as in most expatriate Chinese communities in Asia, they've suffered their share of discrimination in times of nationalist fervor. During the Vietnam War, many were regarded as profiteers, and they came under a certain amount of repression after 1975. Many of them fled Vietnam, making up a large percentage of the boat people who made it to Hong Kong. However, their value to Vietnam's economic revival has put them back in favor today: it's estimated that for every dollar in foreign investment that comes into

investor revived the weekly races in 1991. But don't expect anything as grand as Kentucky or Ascot: the horses look underfed, the jockeys are preteen boys, and the race-goers are usually the city's poorer people — laborers and cyclo drivers. However, it's a quirky otherworldly sort of entertainment and certainly offers a different day, and if you decide on a flutter of your own, the bets are generally just a few hundred *dong*. Races are held every Saturday afternoon, and the track is located north of Cholon near Ho Ky Hoa Park.

Vietnam, two dollars is brought in unofficially by the people of Cholon through family contacts in Hong Kong, Taiwan, and other Asian countries.

Hung Vuong and **Chau Van Liem Boulevards** are the central thoroughfares of Cholon, and the district's most crowded spots are the recently-renovated and colorful **Binh Tay Market** and the nearby long-distance bus station. The district has a great many Chinese restaurants, some of which could be worth visiting. The tumult of Cholon makes a great day trip by cyclo, taxi, or, if you are fearless, a motorbike taxi. The old racetrack and the bare, concrete, time-worn grandstand of **Saigon Racecourse** have sprung back to life since a Hong Kong

WHERE TO STAY

Saigon brims with hotels of all classes, but in Saigon, an international class hotel is far preferable, if at all possible — a retreat from the noise and bustle outside when it can all get to be too much.

Taking it from the top, the bright and glitzy **New World** ((8) 822 8888 FAX (8) 823 0710, 76 Le Lai Street, enjoys a prominent position close to the Ben Thanh Market, with rooms from US$195 to US$850. Over in District Four, the **Equatorial Hotel** ((8) 839 0000 FAX (8) 839 0011 is another pleasant property, with excellent facilities and rooms from US$80 to US$215. The still up-and-coming **Ramada** occupies a magnificent position overlooking the Saigon River — if it is ever finished, it should be extremely pleasant. A **Hyatt Hotel** too, is on the way. Halfway to the airport is the **Omni Saigon** ((8) 844 9222 FAX (8) 845 5234, which, in spite of its distant location, is a quality property with extremely enjoyable rooms, service and dining. Another new joint-venture hotel is the **Saigon Prince** ((8) 822 2999 FAX (8) 824 1888 at 63 Nguyen Hue with rooms from US$180 to US$350.

Among the old tried-and-tested institutions, the **Hotel Continental** ((4) 829 9201 FAX (8) 824 1772, at 132-134 Dong Khoi Street, offers nostalgia and comfortably chic rooms for US$105 to US$170, while the newly renovated **Caravelle Hotel** ((8) 234 999 FAX (8) 243 999, 19 Lam Son Street, should be up and running by 1999. Prices were unavailable at the time of writing.

My favorite is still the **Rex Hotel** ((8) 829 6043 FAX (8) 829 6536, 141 Nguyen Hue Street, with its slightly tacky, very Vietnamese ambience and great hospitality, comfortable, unpretentious rooms filled with facilities and marvelously kitsch touches such as the hide lamp shades. The food, too, is good without being haute cuisine and it is just the sort of place that makes guests feel at home. All 600 rooms are often full. Rooms cost from US$69 to US$659. The renovated **Hotel Majestic** ((8) 829 5512 FAX (8) 829 5510, 1 Dong Khoi, has classic and comfortable double rooms priced from US$130 to US$190, but lacks the warmth and quirkiness of the Rex.

One notch below these classics are the other older and gracious hotels — the popular **Palace Hotel** ((8) 829 2860 FAX (8) 824 4230 at 56-64 Nguyen Hue Street charges US$60 to US$140 for a regular double room and more for a corner room; the friendly, renovated **Bong Seng Hotel** ((8) 829 1516 FAX (8) 829 8076 at 117 Dong Khoi Street is a good deal with full facility rooms from US$50 to US$250. The **Huong Sen Hotel** ((8) 829 9400 FAX (8) 829 0916 at 666-70 Dong Khoi Street is also managed by the Bong Sen, with rates from US$38 to US$95; the **Riverside Hotel** ((8) 822 4038 FAX (8) 825 1417, across the road from the Saigon River, has rooms in the US$90 to US$230 range; and, at the bottom end of the comfort category, the **Saigon Hotel** ((8) 829 9734 FAX (8) 829 1466, 45 Dong Du Street, offers first class doubles at US$40.

The newly-renovated **Grand Hotel** ((8) 230 163 FAX (8) 235 781 at 8 Dong Khoi has magnificent architecture and a pleasant granite-cased lobby with rooms from US$50.

WHERE TO EAT

While all the recommended hotels offer excellent Vietnamese, Chinese and Continental restaurants, good food and exciting restaurants abound in Saigon — there is no shortage. Some older ones are already institutions and are not to be missed, places

such as seedy old **Givral Patisserie and Café**, on Dong Khoi Street, that looks across to the Opera is manned by charming old Frenchified Vietnamese retainers who have been there since the 1960s at least. Then there's **La Bibliothèque** ((8) 823 1438 — another Saigon institution and most elegant and evocative French/Vietnamese restaurant which Madame Nguyen Phuoc Dai, a former lawyer and member of the wartime National Assembly, operates in her villa at 84A Nguyen Du Street, close to Notre Dame Cathedral. The dining room itself is set in what was once her library, so you eat surrounded by shelves packed with old law books.

The ornately decorated, neon-lit **Maxim's** ((8) 829 6676 in Dong Khoi Street, just up from the Majestic Hotel, is another must-visit oddity with an overwhelmingly extensive Continental and Chinese menu,

OPPOSITE: Worshippers at the main altar of the Jade Pagoda. ABOVE: The Continental Hotel offers a breath of nostalgia along with comfortably chic rooms.

live pop music, flashy divas and traditional folk musicians.

A whole rash of trendy restaurants have opened recently — there's a new one every week or two. The best mid-price French restaurant with a classic menu is **Augustin** ((8) 829 2941 at 10 Nguyen Thiep. Next door at 6 Nguyen Hue Street is **Globo Café** ((8) 822 8855, a hip eatery that warms up later in the week. For fine Indian dining, **Ashoka** ((8) 823 1372 at 17/10 Le Thanh Ton, District 1, comes recommended, while **Amigo** ((8) 824 1284 is an open grill with South American flavors, at 55 Nguyen Hue Street, next to the Saigon Prince Hotel. **Bibi's** ((8) 829 5783, at 8A/8D Thai Van Lung, District 1, is run by the former chef from Augustin so it too should be good.

For upmarket Vietnamese food try **Lemongrass** ((8) 822 0496, 4 Nguyen Thiep, a classy three-story place with traditional lute players and chic decor, or **Ba Mien** ((8) 844 2096 at 122b Tran Quoc Thao with specialized regional cooking. **Vietnam House** ((8) 829 1623, 93-95 Dong Khoi Street, has gorgeous waiters working amidst Vietnamese antiques and traditional music.

German food can be had in an authentic Bavarian beer hall atmosphere at **Munchner Hofbrauhaus** ((8) 822 5693, 112 Nguyen Hue Street, opposite the Rex Hotel. There is something eminently pleasant about schnitzel and mashed potatoes after weeks on the road. Other specialties include German sausages and sauerkraut.

NIGHTLIFE

With so many choices, deciding where to go at night can be a problem, or could take days of research. Here are some tried-and-tested venues to help you sort out your options: **Q Bar** ((8) 823 5424 behind the Opera is popular with expatriates and visitors alike; as is the **Gecko Bar** ((8) 824 2754, 71/1a Hai Ba Trung; and **Globo Café** ((8) 822 8855 at 6 Nguyen Thiep is good later in the evening. **Café Latin** ((8) 822 6363, 25 Dong Du, is also said to be good. **Saigon Headlines** ((8) 225 014 is worth exploring at 7 Lam Son Square, and of course **Apocalypse Now** ((8) 824 1463 at 2C Thi Sach is a must-visit — at least for a drink.

HOW TO GET THERE

Over 20 international airlines fly to Saigon and the major ones have offices in town. Saigon is also well connected to the rest of Vietnam through a wide network of flights by the reliable Vietnam Airlines, whose main offices are opposite the Rex Hotel. Any major hotel will have a greeting service at the airport, but failing that the taxis are reasonable — expect to pay up to US$10 for the ride into town. Regional flights from Hanoi, Phnom Penh, Vientiane and Bangkok, Singapore, Jakarta, and Kuala Lumpur are readily available. Saigon is also accessible by land from Phnom Penh for those who wish to risk it.

AROUND SAIGON

Two hundred kilometers (124 miles) north of Saigon, Phan Thiet is the exciting new beach resort getaway for Saigon expatriates — a more attractive alternative to Vung Tau to the south. For the traveler, it makes a fine stop-off point for an overnight stay, halfway between Saigon and Nha Trang along the coast road. Apart from a beach, there is also a fishing port nearby, so good seafood is assured, as are plentiful supplies of pungent *nuoc mam* fish sauce from the nearby factories. Four kilometers (two and a half miles) to the northeast of town, near the beach at Pho Hai, is an eighth-century Cham tower, marking the southernmost extent of the Champa kingdom. The triple tower shows a strong Khmer influence.

Several good beaches are located in the vicinity and there is an upmarket chalet-style hotel known as **Phan Thiet Resort**. The address is Km 9 Phu Hai Beach. Novotel is developing a new hotel adjacent to the **Ocean Dunes Golf Club** ((8) 824 3749 FAX (8) 824 3750. The **Hai Duong Resort** ((62) 848 401/2 FAX (62) 848 403 E-MAIL cocobeach@saigon.teltic.com, also known as **Coco Beach**, has 15 upmarket air-conditioned bungalows and villas, water sports facilities, swimming pool and Jacuzzi for US$62 to US$142 at Km 12.5 Mui Ne Bay.

OPPOSITE: The Rex Hotel roof garden TOP, formerly a United States officer's mess. BOTTOM: Rex Hotel reception area.

VUNG TAU

Once a popular seaside resort under its French name, Cap Saint Jacques, Vung Tau is now a popular Vietnamese resort especially crowded on weekends. It is also a special economic zone and a center of the oil exploration industry in south Vietnam. The beaches here are not that good, but Vung Tau makes up for this with a thriving fishing port, a couple of good Buddhist temples, and some fine seafood restaurants. Other than that, it's

huge statue of Jesus Christ, erected in 1974 at the southern tip of the peninsula, is a must see — after you find your way past the beggars posted along the steps — the view from the top is unbeatable.

Vung Tau's prime attraction is the **Niet Ban Tinh Xa temple**, on the waterfront between Front Beach and the statue, which features a fairly spectacular reclining Buddha and a huge bronze bell on its roof. In short, Vung Tau is a welcome break from Saigon, a place for relaxation and people-watching but nothing much to write home

a relaxing day trip out of Saigon, and the 128-km (79-mile) drive takes you through farmland and a number of bustling rural villages. Even better than the road is the air-conditioned hydrofoil that leaves each morning from the jetty on the Saigon River opposite the Majestic Hotel.

A wander right around the Vung Tau peninsula, from **Front Beach** to **Back Beach**, takes in just about everything of any note, including the old Russian Compound where Soviet oil workers were housed until most of their Vietsovpetro leases were transferred to Western companies after the collapse of the Cold War. The fishing fleet moors at Front Beach, which is fringed with tall coconut palms — a marvelously picturesque spot. A

about. Those who want to relax at a beach resort are better off heading north to Phan Thiet or to Nha Trang.

Where to Stay and Eat

The 76-room **Grand Hotel** ((64) 856 469 FAX (64) 856 088 at 26 Quang Trung Street Bai Truoc costs between US$30 and US$55. Rooms have satellite television, air-conditioning, telephone, and refrigerator. The front rooms with pleasant views over palm trees and market stalls to the beach. The 53-room **Royal Hotel** ((64) 859 852 is the refurbished Canadian Hotel, right on the front beach at 48 Quang Trung Street. Rooms range up to US$120, with a business center and tennis court and garden restaurant. The renovated villa housing the **Petro House Hotel** ((64) 852 014 FAX (64) 852 015, 89 Tran Hung Dao, is very attractive with rooms for US$55 to US$195. The **Anoasis Long Hai Resort** ((8) 822 1467 (Saigon) at Long Hai Ba Ria, Vung Tau, has rooms from US$149 to US$179. Dozens of other hotels and guesthouses are scattered around Vung Tau, but most cater to Vietnamese vacationers from Saigon and have pretty basic facilities.

Try the restaurants along the front beach, most of which specialize in seafood. **Biti BBQ** ((64) 856 652 at 124 Long Ha Street has fresh

priesthood and structure based on the Roman Catholic Church. Its most revered symbol is the "divine eye," which may or may not have been borrowed from Tibetan Buddhism, but is found on all Cao Dai temples in Tay Ninh Province and the Mekong Delta. The eye is the focal point of the sect's spectacular **Great Temple at Long Hoa**, four kilometers (two and a half miles) from Tay Ninh City, where extravagantly costumed services are held four times a day, beginning at 6 AM. But whether it's becoming a tourist attraction or not, a Cao Dai service is something to behold

seafood, Western food and great views, and **Fracivam Restaurant** ((64) 856 320 at 150 Ha Long Street has a French menu.

TAY NINH

Notwithstanding the beauty of the delta region, it's to the north of Saigon that you must go to find one of southern Vietnam's most fascinating cultural oddities. And Tay Ninh, located three hours by road to the northwest, right on the Cambodian border, is the center of one of its unique cultural spectacles, the **Cao Dai**. This religious sect founded in 1926 embraces all religions — mixing Christianity, Buddhism, Islam, Confucianism, and even Taoism into its creed, and featuring a

— massed ranks of cardinals, priests and white-robed male and female clergy parade into the vast, pillared, opulently decorated hall of the Great Temple to pray before the altar and divine eye, to the chanting of choirs and the rattle and chop of wooden instruments.

The Cao Dai lay women are friendly, dignified, and only too willing to explain the principles and rituals of Cao Daism to you before the services begin. Tay Ninh itself enjoys an interesting location and contemporary history. With Cambodian territory

Contrast of color and commerce — fleeting boats CENTER on Vung Tau beach flanked by Saigon's Hotel Continental LEFT and RIGHT Vung Tau's Seabreeze Hotel.

bordering it on three sides, and the southern end of the Ho Chi Minh Trail nearby, the tunnels of Cu Chi and beyond were part of the whole Viet Cong network. A key United States Special Forces base was established there to monitor the end of the Ho Chi Minh Trail during the war. It also came under the pressure of Khmer Rouge border attacks before Vietnam invaded its neighbor in 1979. Tay Ninh is another day trip from Saigon, very easy to reach along a highway which passes through delightful rural scenery — it can easily be combined with the Cu Chi Tunnels.

commodate bulky Westerners, and Japanese and Taiwanese tourists firing AK-47 assault rifles on a nearby range. Never mind — the guides are genuine Vietnamese soldiers and this is as close as most of us will get to the realities of this war. If anyone wants to climb down the spider and rat-infested real tunnels nearby, or even down to the third level of the "tourist tunnels," be my guest. I for one am happy to admire the courage and tenacity of the Vietnamese who survived life in these tunnels against all odds. I don't particularly feel the need to experience their horrors.

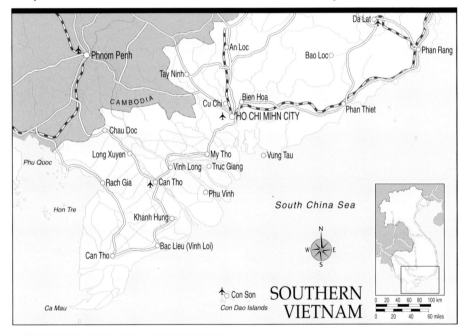

SOUTHERN VIETNAM

South China Sea

CU CHI TUNNELS

The famous tunnels of Cu Chi, the vast **underground network** from which the Viet Cong fought in the Vietnam War, have now taken their place among the great wonders of military lore. They're also one of the prime tourist attractions outside Saigon. But before you go rushing there, remember that you'll see only a little of the more than 200-km (124-mile) labyrinth of tunnels, staging camps, hospitals, operations bunkers and whatever that defied American bombing, defoliation and constant search-and-destroy missions throughout the war.

What you'll see is a small section of tunnels that have been widened slightly to ac-

The Tunnels of Cu Chi by Tom Mangold and John Penycate (Random House, New York, 1985) offers a good insight into life as it was lived beneath the earth's surface, and details many of the ingenious devices employed, not only to stay alive, but also to win their battles.

THE MEKONG DELTA

For some, the Mekong Delta is the heart of Vietnam, a major rice bowl and source of magnificent rural scenery. The Vietnamese

OPPOSITE TOP: A woman rows her sampan to market near Can Tho in the Mekong Delta. BOTTOM: Fishing boats and boatyards on the river at Rach Gia in the Mekong Delta.

call the **Mekong River** *Song Cuu Long* — River of the Nine Dragons — and its vast delta, culminating a 4,500-km (2,800-mile) journey from the Tibetan plateau, is the nation's richest agricultural area. After flowing down through Laos and Cambodia, the river divides into two main arms at Phnom Penh. As they reach the delta, the upper arm flows into the South China Sea at Vinh Long and the lower at Can Tho. In the rainy season from May to October, the entire delta region is virtually a lake, the river's various arms and tributaries flooding the rice fields as far as

suburbs, you pass through rural areas so pretty that it's like driving through a well-kept garden — the rice fields dotted with communities and white stone ancestral tombs. Alternatively, an air-conditioned hydrofoil ferry leaves Saigon at 9 AM each day, departing from a dock at the end of Dong Khoi Street opposite the Majestic Hotel and takes two and a half hours to reach My Tho. Once there, there's not much to see except the exceptionally large and lively **Central Market**, the 100-year-old **Catholic Church** and the **Mekong riverfront**. The **Temple of**

the eye can see. In the dry season it's a dazzling sheet of green or gold, depending on the progress of the crop, punctuated by tiny communities hidden in copses of tall palms.

It's the towns along its banks, and their busy markets, which bring this vast waterway alive — and nowhere is this more evident than in the Mekong Delta. There are a great many interesting delta towns, it all depends on time.

MY THO

This is the closest delta town to Saigon — just 70 km (43 miles) away — and the trip itself is perhaps more interesting than the destination. Right from the edge of Saigon's

the Coconut Monk, located on a small island in the river, is touted as a local attraction, but quite frankly it's not worth the ferry ride — unless you simply want to be able to say you traveled on the mighty Mekong River. My Tho can be enjoyed as a day trip, or as a stopover on a longer tour.

CAN THO

While it takes up to five hours by car to reach Can Tho, the transport of choice is the air-conditioned hydrofoil that leaves from the

OPPOSITE: Young worshippers in the Tay An Pagoda at Chau Doc. ABOVE: Fishing boats at rest in Rach Gia. OVERLEAF: The magnificent Munirangsyaram Pagoda at My Tho.

central pier in Saigon twice a day. The administrative capital of the delta, this bustling and energetic town is quite cosmopolitan — the city was home to thousands of American GIs during the war years. It is located 104 km (65 miles) south of My Tho and is well worth a few days. It makes a good base to explore the surrounding areas. Most visitors come on tours, cramped into the back of a minibus, and I fail to understand why. It is far better to arrive by hydrofoil, then arrange boat trips with some degree of comfort. Enterprising women accost you

on arrival with pictures and glowing recommendations for their tours, and renting a boat is no problem at all. Short tours can be booked from the **Can Tho Tourist Office (** (71) 821 854 FAX (71) 822 719, 20 Hai Ba Trung.

Things to see include the distinctive Khmer Buddhist temple, the **Munirangsyaram Pagoda**, on Hua Binh Boulevard, which is home to a small community of monks. The sprawling **Central Market** in the center of town is a key attraction. The best hotel at present is the expensive 40-room **International Hotel (** (71) 822 079 overlook-

ing the river at 10–12 Hai Ba Trung, with rooms for around US$70. A new international standard **Victoria Hotel Can Tho (**/FAX (71) 829 259 is due to open late 1998. A small hotel with clean and comfortable rooms is the **Ho Guom Hotel (** (71) 827 779 FAX (71) 827 778 at 10 Tru Khoa Huan Street. Rooms cost US$16 to US$20 with air-conditioning, refrigerator, international dial-direct phones and cars or minibuses for rent.

LONG XUYEN

Located upriver to the north of Can Tho, this provincial capital is remarkable for its religious texture. Up until the stirrings of the Vietnam War, it was the center of an armed religious sect called the Hoa Hao, which rejects churches or temples and the priesthood. Obviously, it has left nothing really to show for itself. The city's main showplace, in fact, is its huge **Catholic Church**, completed in 1973 and accommodating 1,000 worshippers. In event of an overnight stay, the **Long Xuyen Hotel (** (76) 841 927 FAX (76) 842 483, 17 Nguyen Van Cung, is regarded as the city's best, with air-conditioned rooms for US$20 to US$25 a night.

CHAU DOC

Lying not far from the Cambodian border, at the confluence of three rivers, this small town is known as a trading transhipment center and pilgrimage center. Just under five kilometers (three miles) west of town is **Sam Mountain**, a famous Buddhist center with many temples and grottoes around its slopes. **Tay An Pagoda** is one of the finest, featuring hundreds of carved wooden images, but there's a more interesting one, the **Cavern Pagoda** halfway up the mountain. It is also notable as the main access point to **Tan Chau**, a famous silk weaving district. The Victoria Hotel Group are constructing a hotel in Chau Doc to be completed early 1999. They plan a luxury boat service from Can Tho to Chau Doc so guests can explore the Delta in style, between the group's two hotels. In the meantime, the very pleasant **Hang Chau Hotel (** (76) 866 196/7 FAX (76) 867 773, 32 Le Loi Street, has rooms with balconies that overlook the river confluence.

ABOVE: The cathedral at My Tho in the Mekong Delta. OPPOSITE: Flamboyant midday mass at the bizarre Cao Dai Temple in Tay Ninh.

THE ISLANDS

Beyond the delta towns, the island of Phu Quoc beckons. See it before it gets too popular—already a small but steady crowd trickles down there and some stay for weeks. Phu Quoc has great potential — a mountainous, forested island in the Gulf of Thailand 15 km (nine miles) south of the Cambodian border, with long unspoiled beaches and coral diving spots as well as a strong military presence which right now occupies the area next to the best beach.

At one time the island held 30,000 to 40,000 prisoners, but presumably they have since left. It just needs the military to be told to lighten up; it is now considered to be a tourist island and heavy-handed soldiers are not good for business. The island is served by Vietnam Airlines with five flights a week from Saigon which land at the main airport in the center of the island near the town of Duong Dong.

Whether you book or not, there will be hotel representatives at the airport to meet the plane and organize transport. A daily ferry runs from Rach Gia to the south of the island to the fishing town of Cay Dua, a very pleasant six-hour voyage. From Cay Dua, it costs around US$5 for a motorbike to take you to one of the resorts or to Duong Dong. If taking the boat, don't forget to bring along snacks, a good book and if possible, take a hammock and get comfortable. I shared the seat in front of the bridge with two other *falang* (foreigners), which we guarded against all comers as we sat playing cards and relaxing.

At the time of writing there are three hotels in Phu Quoc — the concrete government **Huong Bien Hotel** ℂ (77) 846 050 or (77) 846 082 on the outskirts of Duong Dong town overlooks the sea with rooms from US$12 to US$35. Far better are the two chalet-style hotels five kilometers (three miles) south of town. The popular **Kin Linh Hotel** ℂ (77) 846 611 in Duong To village has clean, white, spacious rooms by the beach from US$10 to US$12. The restaurant is on the beach with a sandy floor and hammocks abound. This is basic, no frills relaxation — and it's great. Someone from the hotel meets guests at the

airport. If coming by boat, the port costs 25,000 dong or US$2 and is a half an hour away by motorbike taxi. The third and newest hotel is **Club Tropicana** ℂ (77) 847 127 FAX (77) 847 128, more upmarket and attractive. They can organize fishing and motorbike rental — fine on an almost deserted island. Their rooms range from US$12 to US$25. With fishing the main industry, fresh seafood abounds. Phu Quoc is also the home of some of Vietnam's best fish sauce *nuoc mam*.

CON DAO ISLANDS

With its fine white sand beaches, turquoise waters, coral reefs and forests, this group of 14 islands located 180 km (112 miles) southeast of Vung Tau in the South China Sea is

just beginning to attract tourists after years of notoriety as a very tough penal colony. The main island, **Con Son**, was used as a prison for political dissidents by the French, and then by the United States-backed Saigon government during the war, and the inhumane conditions in which they were incarcerated are illustrated in its **Revolutionary Museum**, on the corner of Le Duan and Ton Duc Thang Streets. You can also visit the former prison buildings.

Home to dozens of varieties of endangered species of birds and sea turtles, the island has recently been conferred national park status. It is linked with Saigon by a Russian helicopter service operated by Vietnam Airlines (round trip costs US$150). The alternative is a 12-hour boat ride from Vung Tau. Two hotels of note on the island are the simple **Phi Yen Hotel** ((64) 830 168 FAX (64) 830 206, 34 Ton Doc Thang, and the Saigon Tourist-operated, quirkily named **Sai Gon-Con Dao Resort** (/FAX (64) 830 155, the spruced up and renovated former residence of the prison wardens, at 18–24 Ton Doc Thang. More expensive, it consists of five villas costing US$30 to US$35 each. The hotel can organize bicycles, cars and motorbikes for rent. **Ho Chi Minh City Heliport** ((64) 884 8814 is at 286 Hoang Hoa Tham, Tan Binh District where you can buy tickets, or at **Fiditourist** ((8) 829 6264 at 71 Dong Khoi where they charge US$5 extra.

The fishing boat harbor on Phu Quoc harbor is awash with color and activity .

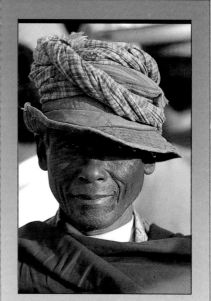

Cambodia

CAMBODIA IS A COUNTRY UNDER A BAD SPELL, the wild card of travel and tourism in Indochina. What is to happen next? With Pol Pot dead and the Khmer Rouge losing its last shreds of power, Cambodia should be about to enter a peaceful new phase, but from all newspaper reports, it seems instead to be slipping further into anarchy. Only time will tell. For Cambodia, the Indochinese War hasn't ended, but continues in different degrees, slipping from one disaster to the next.

Until the tension resolves itself, it requires discretion of anyone planning to travel to Cambodia to study the situation there first. Check your embassy advisory for the latest status. Of course people do manage to travel there all the time. I took the very crowded boat to Siem Reap with dozens of others, from all walks of life, and against prudent advice with no incident, but you can never be sure. There are many good people in the country and the **Ministry of Tourism** ((23) 424 035 FAX (23) 424 036 E-MAIL CIM@Cambodia-Web.net works hard to promote it, but they work against such heartbreakingly high odds.

My advice is linger briefly in Phnom Penh, fly to Siem Reap, explore Angkor Wat and return to the welcoming safety of Laos or Vietnam.

PHNOM PENH

The empty streets of the Cambodian capital make it appear that the end of the world is nigh. Venture out after dark and you're sure it's already here. In the darkened streets, young soldiers (or police, as they are called), armed with AK-47s or worse, stand by in small groups awaiting their chance. Warnings are constantly issued to visitors not to venture out after dark—the locals don't. But head to some of the bars and girly discos and it seems that some folk are willing to risk everything for a little entertainment. A friend who recently returned from Phnom Penh came loaded with tales of kidnapping and intrigue at one of the city's biggest hotels. Then again, another friend returned from a two-week sojourn in the western jungle and reported no problems at all.

What a contrast to a few years earlier when UNTAC were in charge and the streets and businesses were booming. Certainly, soldiers were roaming the streets in droves, but these were well-fed UNTAC soldiers, paid handsomely by the United Nations. The excitement and energy in the air at that time were almost palpable, it was the city of opportunity, journalists and opportunists were there in droves. Restaurants and entertainment spots opened almost daily, catering to the swollen and financially buoyant population; but this has since been replaced by a far tenser and, I should say, more wary atmosphere. Today, instead of foreign entrepreneurs there

are dozens of nongovernmental organizations and young adventurers seeking adventure. The local tourist rag, *Bayon Pearnik*, is indicative of what's happening. Features included a story on National Mines Awareness Day (February 24) and "Keeping Your Head While Losing Your Wallet."

On the other hand, Phnom Penh and Siem Reap have some of Indochina's best hotels — the Hotel Le Royal and the Grand Hotel d'Angkor have been renovated to five-star standard by Singapore's Raffles Group. The Intercontinental has a new property, the

OPPOSITE: Cambodian national flag flies from Foreign Affairs Ministry in Phnom Penh. ABOVE: Traditional music and dance are staged for visitors at the School of Fine Arts near the National Museum.

Sofitel Cambodiana is a major presence, and other hotels like the Royal Phnom Penh are very attractive.

In a sense, Phnom Penh is a real cowboy town — better yet, a speeding vehicle with no one at the wheel. A city whose infrastructure has fallen too far behind to cope. Much of its old French architecture is virtually falling apart before the eyes, its teeming side streets badly potholed and forcing traffic into a state of blaring, elbowing anarchy. With almost every service and utility — water supply, power, sanitation and communications — blowing gaskets trying to keep up with supply.

Like most Asian people, the Cambodians are warm and friendly, although there is understandably a hint of reserve behind their smiles. As some erudite observer noted, "The society is being rebuilt from the bottom up. There are no rules — there are no laws — there is no time for them. The rules will come later when everything's in place." I remember musing over this when I visited one of the city's most popular restaurants, La Paillote on the edge of the Central Market, where a sign on the door read: "Please refrain from smoking marijuana on these premises."

Once it was Le Royal Disco at the Royal Hotel that was packed nightly with young Vietnamese prostitutes, most of them straight out of the rice paddies and tottering and stumbling about in their first high-heels. Now they are found in places such as the well-patronized Martini's Bar, Pub and Disco and a dozen other salubrious night spots around town, and the girls are Khmer, heartbreakingly young and fresh from the country with no ideas about how to protect themselves or demand rights. It is not surprising that AIDS is growing so fast in Cambodia that it could, in terms of deaths, even rival the excesses of Pol Pot and the Khmer Rouge. According to a report from the World Health Organization, tests suggest that at least 100,000 people are already HIV-positive.

The strong and very Asian sense of community that existed before Pol Pot has broken down with the result that people are on their own, with no one to look out for them. An estimated 60,000 commercial sex workers operate in Phnom Penh and sex can be

had for as little as 500 rials, or much less than a dollar. This is the very real tragedy of present day Cambodia, a disintegration of a whole society. When I was there, a little girl fell and hurt herself while playing with her friends. The friends completely ignored her obvious pain and continued to play right in front of her. When I went to see if she was allright, her friends were astonished, laughing at her weakness, although she could barely walk. This brutal lack of feeling is just another legacy of the Khmer Rouge.

GETTING AROUND

Greater Phnom Penh is quite small and a fairly easy place to get around, as it is laid out in a basic grid pattern. If you rent a

motorcycle at US$10 a day (24 hours), you can head where you please.

The city lies north-south along the left bank of the Tonle Sap River at the confluence of the Mekong and the Bassac Rivers — beyond which they all start their long sprawl southwards to finally emerge as the Mekong Delta of southern Vietnam. The most important boulevards, Achar Mean Boulevard and Quai Karl Marx, run north-south, and while the main thoroughfares have names, all other streets are numbered in, it would seem, no particular order. The riverbank and the wide Sisowath Boulevard serve as major points of reference, as does Wat Phnom to the north, the Central Market and the Independence to the south. These streets will provide a reference point for all other areas.

The main north-south street is Mornivong and as it runs south it passes Phnom Penh's derelict railway station, the flamboyantly ugly Central Market and some of the leading hotels — the magnificent Hotel Le Royal, the Pailin, Singapore, Paradise and the Dusit, before turning towards the river at Sihanouk towards the Cambodiana Hotel, the Naga Casino and the Royal Palace.

At first glance the city is derelict, seedy, an uninspiring blend of decrepit French commercial buildings and graceless concrete tenements. Aside from the Royal Palace — which is said to have been designed to rival that of the Thai monarchy in Bangkok —

Archar Mean Boulevard, Phnom Penh's chaotic main street.

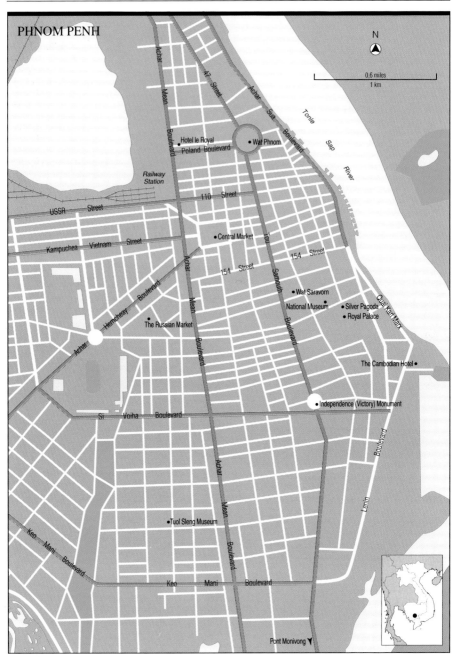

a couple of temples, the Cambodiana Hotel complex and the renovated French villas in the old embassy district, there is really not that much to see. In addition to all of this, along with the remnants of grandeur comes the horror — the dark backdrop of modern-day Phnom Penh — the dreadful relics of "Year Zero."

GENERAL INFORMATION

The government **tourist office** run by the Ministry of Tourism ((23) 426 107 or (23) 427 130 FAX (23) 426 364 or (23) 426 877, at 3 Mornivong Boulevard, dispenses information and tours although a number of private tour companies are also available.

WHAT TO SEE AND DO

Now that the King is often in residence, the opulently decorated **Royal Palace** on Lenin Boulevard between Street 182 and Street 240 is unfortunately not open to visitors except on special occasions such as the King's birthday. Facing a large open park and the river, this inviting edifice, built in 1866 by King Norodom, with its high walls and intriguing glimpses of glittering temple spires and sweeping tiled roofs, is somewhat reminiscent of the Grand Palace in Bangkok, on which it was modeled. Within this alluring complex, the entrance houses an open-air theatre for classical dance performances and an open-air entertainment pavilion for the use of the King. The Royal Throne Room stands by the Throne Hall, crowned with a 59-m (194-ft)-high stupa in a style faintly reminiscent of the towers of the Bayon Temple in Angkor, recalling something of the wealth and splendor of the Khmer civilization. Amid wall and ceiling murals illustrating tales from the *Ramayana*, you'll find several thrones, including a gilded contraption in which the king was carried at royal processions. In the nearby King's Pavilion, watched over by four garudas, there are two more modest sedan chairs which the monarch and his queen presumably used for less official public appearances, rather like using the Toyota to go shopping instead of the Rolls.

One of the more incongruous sights, which can be glimpsed from the adjacent grounds of the silver pagoda, is the **Napoleon III Pavilion**, a complete Normandy-style villa, shipped from France and reassembled in the grounds — a gift from the former French Empress Eugenie early in the twentieth century. Outside the main complex is the compound of the white elephant used for royal occasions, births, deaths, marriages and coronations.

Adjacent to the palace is the magnificent **Silver Pagoda** whose compound is open every morning from 7 AM until noon and from 2 PM until 5 PM. It was one of the few places spared by the Khmer Rouge during the years of madness.

Wat Preah Keo (the **Emerald Buddha Pagoda**) is probably Cambodia's most priceless treasure. Its floor, laid with more than 5,000 solid silver tiles, and the Emerald Buddha itself, is displayed on a dais surrounded by several gold Buddha images, one encrusted with 9,584 diamonds and weighing 94 kg (207 lbs). A huge running frieze illustrating tales from the Khmer version of the *Ramayana*, or *Reamker*, decorates a three-meter (10-ft) wall containing the temple complex, including a paint spattered section that, according to my palace guide,

was ruined by Vietnamese soldiers in their clumsy attempt to renovate (by whitewashing everything), during their stay here (he also noted they were suitably dealt with in the end). Around the pagoda you'll find statues of Cambodia's most recent monarchs, a ceremonial bell tower, various monuments donated by such historical figures as Napoleon III and the small Wat Phnom Mondap, displaying a bronze footprint of the Buddha from Sri Lanka.

Adjacent to palace's northern wall, the **National Museum** is housed in a burnt sienna-colored, classical building constructed in 1917. Open from Tuesday to Sunday, it houses more than 5,000 Khmer paintings and sculptures dating back to the sixth century — to the pre-Angkor states of Funan and Chenla. The collection includes magnificent classical Khmer statues dating up to the thirteenth century. Later treasures include nineteenth-century dance costumes and royal paraphernalia. Just up the street from the

A garish new Central Market building marks the downtown hub of Phnom Penh.

museum, a long string of old shop houses has been transformed into a sort of open college for foreign language students.

One excursion that can only fill a visitor with hope for the future is a visit to the **Fine Arts School** on Rue des Petites Fleurs (Street 70) for the morning dance class which starts at 7 AM (See WITNESS A CULTURAL REVIVAL, page 21 in TOP SPOTS). Here in a crumbling colonial-style complex of the university is a huge barn-like room where hundreds of talented youngsters are being schooled in Khmer classical dance.

As with other Buddhist cities, Phnom Penh is busy with wats and pagodas, and although many were decimated by the Khmer Rouge some have been since rejuvenated. **Wat Phnom** is Phnom Penh's symbol — a distinctly historical place. Located on an artificial hillock north of the city center, it was built in 1434 to commemorate a woman named Penh who reputedly discovered four sacred Buddha images washed there by the river. Hence Phnom Penh — the "Hill of Penh." This crumbling old stupa and surrounding shrines and pavilions were damaged by the Khmer Rouge, and later restored but unfortunately have recently undergone a major restoration, which means clothing the delightful crumbling stupas and statues in cement and brightly colored hues of paint. While it now looks new, rather than 600 years old, it is far less romantic; but still, worth a visit.

ABOVE: Cyclo driver LEFT on Phnom Penh's Lenin Boulevard. Elaborate bas-relief RIGHT at Wat Phnom in Phnom Penh. OPPOSITE: Main gate at the Royal Palace, official residence of King Norodom Sihanouk.

One of the wats with the most atmosphere is **Wat Botum**, a gleaming white construction that houses a monks' residence as part of the complex. **Wat Saravorn**, diagonally opposite the National Museum, is a personal favorite. Old, crumbling and very atmospheric it has the kind of atmosphere that you'd expect from a wat, although lacking any grand architecture. It is also the place for a "Seeing Hands" massage, an nongovernmental organization-sponsored deal where blind men and women massage you through a pair of cotton pajamas and a sheet

which they give you. This is a little too much covering when the ambient temperature is around 35°C (95°F) in the shade.

Wat Ounalom, on Preah Ang Eng, was literally torn apart during Year Zero and has since been restored. It features a stupa which is said to contain a hair from an eyebrow of Buddha, and a marble Buddha image which was smashed to pieces by the Khmer Rouge; but while this is an important Buddhist institute, it has no historical significance: it was built as recently as 1952. **Wat Lang Ka**, located near the Victory Monument (built in 1958 to celebrated Cambodian independence) is another temple which was ravaged by the Khmer Rouge but is now restored and has a small community of monks back in residence.

Wat Koh, situated off Monivong Boulevard at Street 178, and Wat Moha Montrei, which you'll find near the National Sports Stadium, are also worth a visit.

Marking one of the city's major intersections, the distinctively Khmer Victory Monument or Mixay Monument on Sivutha Boulevard was once an independence monument to commemorate the end of French rule in 1953. Although designed in a modern style it features repetitive *naga* or snake motifs, a motif that permeates Khmer art and culture, both historical and modern. The Victory

Monument is now a memorial to servicemen who lost their lives in the Vietnamese-led liberation of Cambodia in 1979, and wreaths are laid at its base on national holidays.

Phnom Penh has several markets, the most imposing of which is Central Market or New Market (Psar Thmei) close to the Wat Phnom — a major, imposing eyesore. This domed, mustard yellow, revolutionary-structured and styled concrete conglomeration resembles a mosque — a fantasy without minarets, that defies any architectural concept of taste. Within the cavernous main hall are brightly-lit displays of glittering

jewelry, gems (including fakes from Thailand) and watches. Surrounding the main building are much more interesting lanes filled with open stalls selling everything from vegetables to piles of dazzling handloomed silks. It is the place to pick up some handloomed *krama*, the ubiquitous Khmer scarves that double as sarong, sunshade or towel. They come in silk and cotton — watch out for the polyester — and make great souvenirs and gifts for just a few dollars. It is also the place to find the ubiquitous tee shirts and pirated video tapes and music cassettes. It's the center of town, and makes a good reference point for urban navigation. Beyond the market are textile stores, restaurants (including La Paillote) and "black market" currency exchange dealers.

More welcoming and far more interesting is Psar Tuol Tom Pong Market or The Russian Market, an enticing, dimly-lit rabbit warren of open stalls selling vegetables, food, and clothes (with great deals in very wearable handloomed silk tops or shirts, at a fraction of the price tagged in the hotel gift shops). Antiques scoured from the countryside, handloomed silks, tools, ceramics — the market is filled to overflowing with exciting buys. Don't expect to take a few minutes — dedicated browsers can spend hours there — but keep your bag secure, pickpockets are rumored to roam here, although (for once) I had no trouble at all.

The city's most dramatic monuments are those connected with the darkness and brutality of Pol Pot's reign of terror. Tuol Sleng Museum was formerly Tuol Svay Prey High School, which the Khmer Rouge converted into Security Prison 21 — their main detention, interrogation and torture center for class enemies during Year Zero. Within the bleak brick walls on Street 113 is a collection of old school buildings, surrounded by barbed wire, which even now exude an atmosphere of chill and fear. Although a lot of its most hideous features — the various instruments of torture — have been removed, it still chills the heart to think of the thousands of people who passed through this charnel house. Most shocking of all is the vast gallery of small passport-size before- and after-interrogation photos of the detainees condemned to death and torture — with their ages, exhibited on

ABOVE: Mixay Monument forms a major orientation point in Phnom Penh.
OPPOSITE: The spires of Wat Botum.

wall after wall within the main buildings. Most of them were hardly older than teenagers, stunned and completely helpless, in the grip of a regime that was all the more insane for the grim efficiency with which it recorded its purge.

WHERE TO STAY

Phnom Penh is not short of excellent hotels, a lot of them constructed or renovated during the UNTAC years. Even the smaller properties have fax and telephone services, CNN/

who stay here, but rather the high-flying yuppies who haunt Asia. Room rates start around US$280 a night.

The **Sofitel Cambodiana Hotel** ((23) 426 288 FAX (23) 426 392, on the southern reaches of Quai Karl Marx, has a main wing of 360 rooms surrounded by restaurants, bars, shops, garden bistros, the best book shop in town, and an excellent gift shop all set on the bank of the Bassac River. It's a grand palace of a hotel, which took from 1967 to 1987 to build, the construction halted for years by warfare and the Khmer Rouge

Star-Plus satellite services either in the lobby or rooms, a reasonable breakfast, and rates in the US$30 to US$45 range.

The best address is the **Hotel Le Royal** ((23) 981 888 FAX (23) 981 168 E-MAIL raffles .hir.ghda@bigpond.com.kh, 92 Rukak Vithei Daun Penh off Monivong Boulevard, which looks as impressive from the outside, as it does within. Its imposing, ornate French façade, and the role it played during the Khmer Rouge takeover, gives it a certain mystique. The beautiful, if chilly, renovated rooms and public areas are just as grand as the day it opened in the 1930s. Refurbished and run by the Singapore Raffles' Group, the price is high and the ambience is as expected. Today it is not the elegant world travelers

purges. This is a fine place to stay if you want absolute comfort and all facilities, but expect to pay around US$200 a night.

The **Hotel Intercontinental Phnom Penh** ((23) 720 888 FAX (23) 720 885, 296 Boulevard Mao Tse Tung, is another recent addition to Phnom Penh's hotel scene that comes replete with air-conditioning and every facility, all contained in a comfortable and gilded package. Rooms cost from US$170 to US$1500 per night but no doubt discounts of 20 to 23% are available for this and all the big hotels. Another very attractive hotel is the **Royal Phnom Penh** ((23) 360 026 FAX (23) 360 036 at Samdech Sotheroh Boulevard, Sankat Ton Le Bassac with 75 rooms in a garden setting by the Bassac River. The hotel has a pool and a driving range.

For a much more satisfying taste of old Indochina, at a fraction of the price, the **Renaske Hotel**, opposite the southern end of the Royal Palace complex on Samdech Sotheros Street and just a few blocks away from the Cambodiana, is highly recommended. This former Buddhist institute has comfortable rooms with satellite television, hot water and air-conditioning and charges US$35 to US$40 a night. The charming open, tiled verandah and lobby look out on to gardens of frangipani trees and, on the right night, the moon rising over the temple spires

place for a drink at sunset, overlooking the river, but the food is of the best in town and the drinks are excellent.

The Thai-owned **La Paillote** at 234 Street 130 in the La Paillotte Hotel has excellent French and Continental cuisine, plus Thai favorites, at very pleasant prices. As for Thai food in an upmarket setting, the **Chao Praya**, set up in another restored villa at 67 Norodom Boulevard serves buffets and is popular with tour groups. **Apsara** at 361 Sisowath Quay serves Khmer food and **Saigon House** at 212 Sisowath Quay serves Vietnamese food.

of the Royal Palace. The food is fairly basic — a continental breakfast is included in the rate and other meals can be taken elsewhere. The coffee is good, and staff friendly and accommodating.

WHERE TO EAT

Today's Phnom Penh is a far cry from the days in the early 1990s when UNTAC personnel from 39 countries created a wild and raucous cowboy boomtown. Today it has reverted to a town in waiting, and many of the restaurants no longer exist. One that does still exist, and in fine style, is the **Foreign Correspondent's Club** (FCC) ((23) 427 757, 363 Sisowath Boulevard. Not only is it a fine

For a feast of Khmer specialties or continental food try **Le Royal Restaurant** ((23) 981 888 at Le Royal Hotel.

HOW TO GET THERE

Phnom Penh is serviced with daily flights from Saigon and Vientiane on the national airlines as well as direct flights from Singapore, Kuala Lumpur and Bangkok. The drive from the airport to the city is short, costing only a few dollars, although the bigger hotels

OPPOSITE: The No Problem and Mousson pub and restaurant in Phnom Penh. ABOVE: Hand-woven silks and other traditional textiles can be found in shops ringing the Central Market. OVERLEAF: Girls in traditional dress in Siam Reap.

pick up their guests. Adventurers wanting to save money take the bus from Saigon which is the least expensive, if not the most safe, method of getting there.

AROUND PHNOM PENH

From Tuol Sleng, a visit leads inevitably to the Killing Fields themselves — or rather, to one of the many mass-execution spots that operated throughout Cambodia during the Khmer Rouge reign. You get to the former extermination camp of **Choeung Ek**

past tranquil *padi* fields and small farming communities. This trip is a perverse contrast of rural industry, happy-go-lucky children who shout "Hello-good-bye!" as you ride past, haughty russet-colored Brahmin cattle hauling traditional rice carts, and the horror that awaits you at Choeung Ek. There, about 15 km (nine miles) from the city, a tall glass-faced monument contains nearly 9,000 skulls, the remains of victims dug from mass graves nearby. The grim relics are arranged in tier upon tier up the tower according to sex and age. No amount of description can really prepare anyone for these monuments, and it leaves us all a little humbled by the contrasting good fortune of our own lives.

PHNOM UDONG

One of the journeys to make beyond Phnom Penh takes you to a hilltop complex of temples and stupas at Udong, about 45 km (28 miles) north of the city on Road No. 5, which was the capital of Cambodia from

1618 to 1866. Tour agencies will take you there by car, or you can get there by motorbike—but either way, it's a long, hard drive. While little remains of King Ang Duong's masterpiece — the main temple of **Vihear Preah Chaul Nipean**, almost 100 pagodas and its main relic, a huge sitting Buddha, were blown to pieces by the Khmer Rouge, three huge stupas nearby were mercifully spared. Decorated with *garudas*, elephants and ceramic motifs, they stand untouched, even though they commemorate three of the country's kings. From these stupas you

look across a vast rice plain, contemplating once again the chilling juxtaposition of darkness and light in Cambodia. Below the hill there's a **memorial to victims** recovered from more than 100 mass graves in the region. Udong is a popular pilgrimage site where Khmers go to pay homage to former kings.

TA PROMH TONLE BATI

About 35 km (23 miles) south of Phnom Penh on Road No. 2, is a twelfth-century temple dedicated to both Buddhism and Brahmanism. Built by the Khmer King Jayavarman VII this temple can be combined in a day visit with Phom Chiso.

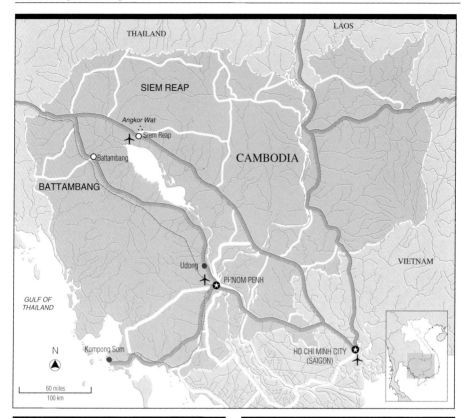

PHOM CHISO (SURYA PARVATA)

Also south on Road No. 2, this eleventh-century Angkor mountain temple stands atop a 100-m (328-ft) hilltop with extensive views over the surrounding countryside. Built by King Suryavarman I, this laterite, brick and sandstone monument has a central sanctuary housing the "Black Lady" or Neang Khmao, a venerated Buddha about 300 years old.

KOH DACH

Reachable by Road No. 6A about 15 km (10 miles) northeast of Phnom Penh, or by boat along the Mekong, past fishing villages and idyllic rural scenes, Koh Dach is a Mekong island of about 30 sq. km (12 sq. miles), an historical center well known among the Khmers for its quality handloom silk weaving. It is also a center for pottery production, wood carving, painting and gem cutting, making an interesting day's outing from Phnom Penh.

SIHANOUKVILLE (KOMPONG SOM)

This town on the south coast, bordering the Gulf of Siam, is about 230 km (160 miles) from Phnom Penh by car, or a 25-minute flight from Phnom Penh (Royal Air Camboge charter is available on request). Cambodia's main port and a major fishing town, Sihanoukville is struggling to regain its importance after so much strife. While it is touted as a resort town with superb beaches, with several hotels and restaurants and a few brave souls working to resurrect a tourism industry along with scuba diving, it is wise to check the current status before heading there. For now, a more accurate description is that Kompong Som may one day become an important access to virtually untouched islands, and their resort potential, that lie off the coast. In the meantime, Angkor Wat looks good.

What's perhaps more off-putting is that the road also takes you through one of

OPPOSITE: Skulls pack the macabre shrine to Pol Pot's victims at Choeung Ek LEFT. Sun sets over the Mekong River RIGHT in Phnom Penh.

Cambodia's prime logging areas, with depots and sawmills lining its sides for many kilometers, giving a graphic picture of the extent to which the country's forests are being raped.

Stay at the **Seaside Hotel** (/FAX (34) 345 523, opposite the beach at Mondul 1 Sangkat 4, or at **Hawaii Hotel** ((34) 933 447 FAX (34) 340 267, Mondul 2 Sangkat 2.

ANGKOR WAT (SIEM REAP)

This vast, world famous complex of Hindu-Khmer temples, the legacy of the Khmer

Empire, has been stirring the imagination of visitors for more than a century. Created over the ninth through the fifteenth centuries, Angkor has been compared to Egypt's pyramids as a world-class monument, for its scale of vision and the extraordinary artistry of both the architecture and the quality of the carving. Angkor is what draws visitors to Cambodia, and to consider coming to Indochina without visiting Angkor is almost unthinkable.

When I was there in early 1998, things were looking very positive. Teams of Japanese archeological experts were working on a restoration project of the Bayon. A short 300-m (200-yard) walk revealed the French-restored **Elephant Terrace** and **Terrace of**

the **Leper King**, where a replica statue of the Leper King stands guard over what was possibly the royal crematorium. Nearby, in a tree-shaded grove, a long, stone causeway led to the mountain temple **Baphuon** where ancient masonry lay about in ordered piles, awaiting rebuilding by a French team to be completed by the year 2003. Within the complex, the late tenth-century **Phimeanakas** was also being restored, due for completion in 1999. The beautiful **Banteay Srei** Complex and **Banteay Samre** have been opened to visitors, with the Khmer Rouge threat re-

duced in those areas to the north, and only one minor checkpoint blocking the road, which was in the process of being upgraded. A much larger part of Angkor is now open to exploration although its three most famous attractions continue to be Angkor Wat, the Bayon Temple with the Baphuon and Ta Phrom.

In general tour guides, which you find in Siem Reap, the main access point to the ruins, are very accommodating, guiding without hustling their charges from one complex

OPPOSITE: The South Gate at Angkor Wat. ABOVE: Many artistic treasures, such as these incredible bas-reliefs, have survived Cambodia's years of conflict. OVERLEAF: Stone causeway leads to the main entrance and architectural wonders of Angkor Wat.

to another, content to spend as much time as desired at each site — as long as you're ready to head back to town by nightfall. All hotels have guides available, or if you prefer, try **Angkor Tourism** ((23) 57466 FAX (23) 57693, on Street 6, Phum Sala Kanseng, Siam Reap.

THE TEMPLE TOUR

What you see in Angkor today is only a fraction of its ancient splendor, a small part of what was once a powerful empire whose

on to a vast courtyard that leads to the main temple itself. This gigantic three-story structure, rising 55 m (180 ft) from the ground, was built by King Suryavarman II (1112–1152) and dedicated to Vishnu. Of all the monuments of Angkor, it is this great structure, quite apart from the wealth of bas-relief carvings of Hindu epics and Khmer battles adorning the walls and cloisters surrounding it, which testifies to the power and glory of the Khmer civilization.

By contrast, the **Temple of Bayon**, is the most fantastic and evocative sight of all of

influence and architecture spread across Cambodia, into southern Laos and Vietnam and towards Yunnan in southern China. Even so, **Angkor Wat** itself, the first stop on the temple tour, continues to awe visitors with the monumental vision and artistry employed in its construction. You approach the outer temple walls along a wide, romantically dilapidated stone causeway that runs across what was once a protective moat. The main gate and its walls are spectacular enough, crowded with sculptures of *apsaras* and Hindu carvings, but this in turn opens

Angkor. This awesome chunk of masonry rising through the jungle with its beautiful implacable faces carved on every side was built to create awe in the beholder, which it does, like an image from an Indiana Jones-type movie. Even though most is now in ruins (a Japanese team is busy restoring sections at the time of writing), this temple generates such a power, it's almost tangible.

Built in the reign of Jayavarman VII (1181–1201), it lies in a forest clearing. Coming from Angkor Wat, you enter by the southern gate along a causeway lined with 54 statues, many of them now headless. To the left, demons and to the right, gods, hold seven-headed snakes or *nagas*, a much used Khmer symbol. Once the epicenter of the city of **Angkor**

ABOVE: Angkor Thom is at its most atmospheric in the early morning. OPPOSITE: Angkor's main surviving ruins are still in the crossfire of government and Khmer Rouge troops.

Thom, the Bayon stands at the symbolic center of heaven and earth, and five entrance gates are placed at strategic points around the complex.

Where Angkor Wat sheds an almost golden glow in the late afternoon sunlight, Bayon is best visited in the morning when the rays of sun slanting through the trees light up its mass and somber tones. Granite gray, the stone of its walls, towers and stairways crumbled and pitted by time; its artistry arouses a sense of majesty and mystery — its upper level mounted by no less

than 49 towers bearing 172 huge, Buddhalike faces are thought to be carved in the likeness of Jayavarman VII in the form of the Bodhisattva Avalokitesvara. Wherever you stand, these enigmatic, implacable features are watching you, some of them so big that they fill your view, others barely recognizable within the stonework, so harshly have they been worn by elements. Elsewhere in the galleries and cloisters of this momentous relic, some 11,000 other carvings depict Hindu legends and Khmer life at that time.

Adjacent to the Bayon to the east, is the 1,150 ft (350 m) **Terrace of the Elephants** which was used as a dais from where Khmer rulers reviewed their armies. From one end

to the other, it is sculpted with bas-relief cavalry and war elephants. This terrace leads to the **Terrace of the Leper King**, at present under renovation. A replica sandstone statue of the Leper King can be seen within this complex.

Ta Phrom is the most derelict of all the Angkor temples, but it is also the relic that most inspires the sense of intrepid discovery, as if you are the first to arrive on an undiscovered site, just as much of the complex once was before decades of restorations. Built as a Buddhist temple in the

late Angkor period, by Jayavarman VII, from the eleventh to the beginning of the thirteenth centuries, it has since been virtually reclaimed by the jungle, its stonework mingled with the spreading roots of Banyan trees, broken and heaved about by their tree roots, as though an earthquake had struck it, and its passageways and cloisters piled with fallen masonry.

Elsewhere, the impressive **Ta Keo**, built by Jayavarman V (968–1001) in honor of Shiva, rises up out of the jungle in a series of lofty towers with huge stone steps that are exhilarating to climb.

Huge Buddha face OPPOSITE at the awesome Bayon Temple alongside the bas-reliefs ABOVE that adorn the walls and niches.

In the forest around the Bayon, huge **ceremonial gates**, all that remain of the precincts of Angkor Thom, have approaches lined with rows of sculptured guards — most of them now headless after centuries of looting. In attempt to halt the destruction and pillage at Angkor, the Angkor Conservation Agency has managed to save about 5,000 statues and other relics, storing them in sheds at its compound between Siem Reap and Angkor Wat.

BANTEAY SREI AND BANTEAY SAMRE

Further out, about 32 km (20 miles) from Angkor, the newly opened **Banteay Srei**, or the **Citadel of Women**, is one of the oldest Angkor monuments and one of the finest in execution. Built by King Jayavarman V in the second half of the tenth century and dedicated in AD 987, to Shiva, the pink sandstone temple is small and delicate, filled with the finest flutings and carvings of apsaras and divinities. At one time the entrance path was lined with linga, although many have since collapsed or fallen into disrepair. The distance from Angkor deters many visitors, which makes it mercifully quieter than the main complexes. It is best visited in the cool of the early morning or late afternoon.

Also newly opened, about 18 km (11 miles) from Angkor located at a turnoff before Banteay Srei, **Banteay Samre** is a twelfth-century complex of some interest and in quite reasonable condition. Its central temple is flanked by two libraries and crumbling wings surrounded by what was no doubt a moat. There are two main entrances in the high, slightly forbidding walls that surround the complex. Amidst the fallen lintels and monumental architecture, a group of feral, Indian-looking children — were they the ancestors of some long ago craftsmen? — swarmed over the ruins like a tribe of wild monkeys.

A dedicated visitor could easily spend a week or even two exploring the ruins, but most visitors will find two or three days is sufficient to get a feel of this marvelous place.

WHERE TO STAY

Siem Reap, the main access point to the Angkor ruins, is a fairly idyllic little market town, dissected by a small river. It also has one of Cambodia's most famous hotels, the **Grand Hotel d'Angkor** ((63) 963 888 FAX (63) 963 168, 1 Vithei Charles de Gaulle, recently renovated, revitalized (and repriced) by Singapore's Raffles Group. It includes five restaurants and a spa. Outside the Grand, several new hotels and guesthouses have sprung up to compete for tourists. The Grand operates one of them, the **Villa Apsara**, right across the street from the main hotel, with a pool and chalet-style rooms at US$55 a night.

The **Angkor Hotel** ((63) 380 027 FAX (63) 380 027, on Road No. 6 to the Airport, opening in 1998, has 111 rooms and all facilities. Another new property, **Golden Apsara Guesthouse**, 220 Mondol 1, has rooms for US$10 to US$20, while the unfortunately-named **Hotel Stung** on Wath Prom Rath Street charges US$50 to US$80. The **Hôtel de la Paix** on Sivatha Street, Sangkat 11, was an empty shell when I was there but no doubt it is operating again by now. The **Green Garden Home Guest House** ((15) 631 364, 51 Sivutha Boulevard, is a pleasant place with eight rooms from US$8 to US$17, while the large clean and efficient **Freedom Hotel** ((63) 963 473 FAX (63) 963 473, on Road No. 6 just across the bridge near the turn off to Angkor, is well run if lacking all atmosphere, with rooms from US$15 to US$30.

WHERE TO EAT

Most visitors eat in their hotels, but several good restaurants and bars await discovery for those not on an all-inclusive package. Places to try include the **Restaurant Samapheap**, with an open dining pavilion and garden, located just across the river from the Grand Hotel d'Angkor. The **Green House**, at 58 Mondol Street on the road in

from the airport, is a new Thai-style restaurant and guest house, in a big pavilion with a wide range of dishes from Thai to Khmer, to Western and Chinese. The **Bayon**, on the riverside road in Siem Reap, is a popular stop for Khmer, and other Asian dishes. The **Royal Restaurant** ((63) 964 143 serves Khmer, Thai, Chinese and Western specialties in a outdoor eatery at 55 Sisowath Street. Within the Grand Hotel, visitors can treat themselves to the best in quality and service at the **Café D'Angkor**.

HOW TO GET THERE

Siem Reap can be reached from Phnom Penh by twice-daily flights on Royal Air Camboge.

More flights are scheduled during busy times, so check with their office in Phnom Penh ((23) 428 891 FAX (23) 428 895, 62 Tou Samouth Boulevard.

An interesting alternative is the fast, crowded, and not very comfortable express boat that takes about four hours. From August to October, when the river is high, more boats are available and the trip is no doubt less torturous. Depending on the political situation when you go, it may or may not be safe. It was not recommended when I was there, but the morning I took the boat, it was

filled to the brim with backpackers and the odd adventurer; the journey was completed without incident. Boat offices are all located north of town along the Bassac River. When I left, I took the (very welcome) flight back to Phnom Penh.

OPPOSITE: Siem Reap's Grand Hotel, gateway to Angkor Wat. ABOVE: Newlyweds LEFT in Phnom Penh testify to Cambodia's arduous recovery from horror. Girl in traditional dress RIGHT at Siem Reap.

Laos

THIS SLEEPY, MOUNTAINOUS DOMAIN IS THE JEWEL of Indochina, a golden land where the charm of the people is largely untainted, where saffron-clad monks chant prayers on their morning alms round, where religion and graciousness are more important than money. How long can it last? Probably not long, but while it does, it is something to treasure and now is the best time to visit this gorgeous country.

Laos is not so much a land of breathtaking sights and monumental edifices, but rather a land of charm, a place to spend qual-

ity time. Best of all, Laos allows for a wonderful feeling of adventure, of striking new paths rather than following the tried and tested routes of thousands before you. To the north are untrammeled destinations which are home to hill-tribe people following the lives of their ancestors. Their lives are regulated by harvest festivals tied to the movements of the moon rather than calendar dates, and the details of their traditional dress should be perfect in execution.

There are so many navigable rivers to explore, including stretches of the Mekong where tourists, and even some groups, are making their way down from the northern Thai border near Huay Xai to the ancient royal capital of Luang Prabang — a two-day jour-

ney, whether you take a clattering old river transport or fast new speedboat, that passes through rugged mountains and authentic tribal villages. Where Vietnam faces dramatic economic and industrial development, with all the environmental problems that it will inevitably bring, and Cambodia struggles with its divisive political crisis, the Lao government struggles with laws to conserve their country's riches — "step by step" is their motto, no headlong rush into uncontrolled development just for the sake of a quick money fix.

Laos promises several years yet of rustic serenity — a relatively mystical experience within the continuing upheaval of Indochina. So much of this buffer state is undeveloped, comparatively undiscovered, awaiting exploration. Tourists are still guests, rather than merely walking dollar signs as can often be the case in some neighboring countries. So much of its French colonial character is still there, gracing its main cities and towns. So much of its spiritual life is intact, the Buddhist temples are flourishing, its festivals reaffirming and renewing faith to the gentle culture, its people reaching back to traditions that have barely acknowledged nearly two decades of Communist rule. Even the Australian-funded Friendship Bridge over the Mekong River, which links the country directly with Thailand, is little more than a symbolic step into modern Asia, and although there are more of the ubiquitous Thai *tuk tuks* and *samlor* motor transports on the roads, there is no obvious negative impact from this convenient new access.

That's not to say that Laos won't change. The country remains threatened by the struggle between Thailand and Vietnam for economic and political dominance. Its relative poverty — poverty, that is, in relation to the industrial and trading growth of the rest of Southeast Asia — means that an industrial base of some sort must be developed soon if it is to compete with the rest of the region; but the Lao people are taking their time. For now, to travel in Laos is to travel back to the more graceful days of Indochina and to experience at first hand the gracious

OPPOSITE: Courtyard of Vientiane's Pha That Luang (Great Sacred Stupa). ABOVE: The stupa soars behind a statue of King Setthathirat.

Laos   227

Buddhist traditions which have been corrupted by rapid economic growth and too much tourism in neighboring Thailand.

As elsewhere in this region, Laos has two seasons, governed by the monsoons that bring rain and high humidity during the summer months from May to October and almost perfect conditions — warm and dry in the daytime, cool and dry at night — from November to April. However, temperatures get more frigid, especially at night, in the northern mountains during the dry season, bringing cold relief for those from tropical

countries. You'll find that United States dollars and Thai *baht* are accepted all over Laos as compatible currency with the local *kip*. Major credit cards are also accepted. Laos, or at least Vientiane, has the best and cheapest international telecommunications in Indochina, with calls costing a fraction of what they do in Vietnam and Cambodia.

Laos is a place where, once you're there, the rest of the world just doesn't seem to matter any more.

The whole of Laos is to become accessible for Visit Laos Year 1999, making previously

closed areas open to visitors. In the past few years the north of the country has opened up so that now visitors are welcome to make their way up as far as Muang Sing near the China border and to Xieng Kok at the Myanmar border where some travelers cross over into Myanmar. To the east, the previously restricted area of Phong Saly has been opened as has the area to the east of Xieng Kuang, to Sam Neua. To the south, the charmingly untouched town of Attapeu on the Sekong River has become accessible and the tribal highlands beyond which were previously off the list. Land border crossings too, have increased dramatically. Although the northern Vietnamese highlands are still not accessible to foreigners from northern Laos, northern Thailand certainly is, which has resulted in a steady stream of visitors making their way down the Mekong to Luang Prabang.

VIENTIANE

The Vientiane of Paul Theroux's *The Great Railway Bazaar* (1975), which he dismisses so disparagingly as a city where "the brothels are cleaner than the hotels, marijuana is cheaper than pipe tobacco and opium easier to find than a glass of cold beer," has for better or worse, disappeared, to be replaced with more traditional Buddhist values and busy nongovernmental organizations (NGOs) intent on setting the country to rights — a generally terrifying prospect. But then, it still has the character and feel of a large market town with an exotic veneer of French culture, an undercurrent of Buddhist sanctity, a bustle of largely Thai and ethnic-Chinese trade and enterprise, and a tendency to regard a traffic jam as a bank-up of too many bicycles. It has more than bicycles, of course, and in fact an astonishing stream of traffic — jeeps, the ubiquitous white land-cruisers of the innumerable NGOs, minibuses, jumbos and *tuk tuks* imported from Thailand, cars, motorcycles, bicycles and pedal-cyclos — flows all day along the main streets of Samsenthai Road and Lan Xang Road.

GETTING AROUND

The city snuggles along a broad curve of the Mekong, hugging the western bank, and

ABOVE and OPPOSITE: Two views of Wat Si Muang, one of Vientiane's many well-preserved temples. OVERLEAF: Sunset on the Mekong River in Vientiane.

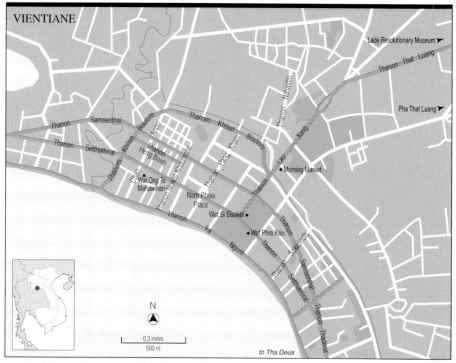

VIENTIANE

Laos Revolutionary Museum ➤

Thanon That Luang

Pha That Luang ➤

Thanon Samsenthai

Thanon Khoun

Thanon

Thanon Setthathirat

Thanon Khoun Boulom

Thanon Arou

Thanon Heng Boun

Thanon Pang Kham

Boulom

Thanon Chao Anou

Wat Ong Te Mahawihan

Thanon Nokeo Koummane

Nam Phou Place

Wat Si Saket •

Thanon Fa

Thanon Nguni

• Wat Phra Kèo

Thanon Nahaidio

Thanon Lan Xang

• Morning Market

Thanon Samsenthai

Thanon Sakarinth

Thanon Setthathirat

Thanon Thatfatsa

N

0,3 miles
500 m

to Tha Deua

most of its main streets, like Samsenthai Road, run parallel to the river. Others radiate to the north, such as Lan Xang Road, which is a wide triumphal boulevard leading to the city's most notable landmark, the towering Pratuxai Monument, which looks as though it were uprooted from somewhere in the vicinity of India's Taj Mahal (word has it that it was built with United States funds donated to extend the airport some years ago). The city's tourist and business district begins in downtown Samsenthai Road, around the intersection with Lan Xang Road and extends southwest to the river, taking in the very 1960s-style Nam Phou Fountain which dominates a circle lined with restaurants, the Laos Tourism Office, the Diethelm Travel Company and the rear entrance to Lan Xang Hotel.

Vientiane is an easy city to deal with — compact, flat, and with most major hotels, restaurants, agencies and attractions within walking distance, or at least in directions you can easily decipher. Only when you head north to locations beyond Pratuxai Monument do you need transportation, which the passing *tuk tuks* handle admirably. If you prefer, a bicycle or motorbike can easily be

rented in Vientiane for a day or week at ridiculously reasonable prices.

GENERAL INFORMATION

Inter Lao Tourism ((21) 214 832, on Setthathirat Road near the Nam Phu Square, provides very good maps of Vientiane and the whole country, while the **National Geographic Department**, adjacent to Le Parasol Blanc just past the Patuxai Monument, sells detailed, large-scale maps for a very reasonable price. Also, the Women's International Group produces the annual *Vientiane Guide*, which is mainly for incoming expatriate families but is also of great help to tourists. You can buy it in most hotels, and the proceeds go to support aid projects for women and children.

WHAT TO SEE AND DO

Standing on the banks of the Mekong, Vientiane has plenty of things to enjoy, like the leisurely downing of a cold beer or an aperitif at one of the riverside cafés at sunset, or a tour of some of the more beautiful wats and

Ornately decorated entrance to the main prayer hall at Wat Si Saket.

a relaxing herbal sauna and massage at Wat Sok Pa Luang. It is also an easy place to spend hours prowling the numerous galleries, the morning market, and souvenir and art shops for beautiful textiles and antiques. Happily Vientiane is not overburdened with too many sights, and a leisurely few hours spent sightseeing should leave you with time to spare for relaxation.

The city does, however, have some beautiful temples and wats which are worth a wander. The most important is dominated by the massive golden stupa of **Pha That**

elephant is led to the stupa while ritual processions celebrate the event at temples all over the city. Hundreds of monks and novices pour in from the provinces to take up a vigil in the cloisters around the stupa. At dawn on the second day, they're presented with alms and new robes. On the third day, thousands of faithful Buddhists walk to the stupa from the other temples, led by musicians and bearing offerings. Chanting and praying, they circle the stupa until the late hours of the night, many stopping to light candles and pray amidst the lively crowds.

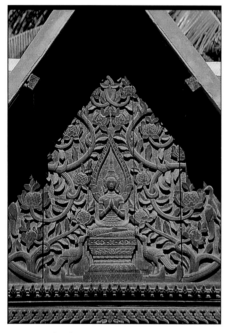

Luang, which towers over two monasteries and walled cloisters four kilometers (two and a half miles) north of the city center at the end of That Luang Street. The most revered Buddhist monument in Laos, That Luang was built on the site of a Khmer temple by King Setthathirat in the mid-sixteenth century when Vientiane became the capital of the kingdom of Lan Xang. A statue of the founder sits atop a stone column in an enclosure in front of the huge timber main gates. That Luang was virtually destroyed in 1828 by invading Thais, and the present structure is the result of a French restoration in 1900.

In November each year, the complex becomes the focus of Vientiane's biggest religious festival. On the first day, a sacred white

Wat Si Saket is one of Vientiane's more splendid monasteries, and its oldest (it hasn't had to be rebuilt as have many of the older ones), located on a corner of the intersection of Lan Xang Road and Setthathirat Road. It was built in 1818 and its architecture reveals a strong Thai influence, while the inner walls of the compound are packed with Laos-style Buddha images. A Khmer image of Buddha seated under a canopy formed by a multi-headed cobra is one the hall's showpieces, along with statues that were damaged in the war. The wat's main hall, also Thai-style, is lined with Buddhist murals and its ceiling is decorated with a floral design similar to those in the Thai temples of Ayutthaya. **Wat Pha Kaew**, just opposite Wat Si Saket, was rebuilt in 1942 on the site of the former royal temple of the Lao kings. It's now a museum, with terraces and a main hall crowded with impressive Buddha images dating back to the sixth century, along with a gilded throne and, in a garden to the rear of the hall, a stone jar from the Plain of Jars.

Wat Ong Teu Mahawihan, on Setthathirat Road near Chao Anou Road, was also built in the sixteenth century, about the same time as That Luang, and was similarly destroyed in wars with the Siamese and later rebuilt. The name translates as "Temple of the Heavy

Wat Sok Pa Luang and nearby Wat Si Amphon, south of the city center near the Mongolian Embassy, both offer something more than the usual temple tour. They provide soothing herbal saunas, administered by nuns, and therapeutic massage if you require it. Although there is no charge, a donation of a few dollars is appreciated.

The architecture of the old French mansion in which the Laos Revolutionary Museum is housed on Samsenthai Road is one good reason for a visit. The museum is devoted to depictions of the Pathet Lao's

Buddha," and it is dedicated to a massive sixteenth-century Buddha image made of bronze in its exquisitely decorated main hall. It is one of the most important wats in all of Laos. Wat Si Muang, east of the city center at the confluence of Setthathirat and Samsenthai Roads, was built in 1566 and houses the foundation pillar of the city. The story goes that during the construction of the temple, a pregnant woman named Nang Si was inspired to sacrifice her life and jumped into the hole to be crushed by the pillar. She became known as the protector of Vientiane and is a revered inhabitant with people coming to pay her homage. Wat Si Muang is the site of a huge procession two days before the That Luang Festival in November.

struggle for power, with pictures and some historic weapons. An exhibition of Khmer sculptures and Lao musical instruments can also be seen. The museum opens, sometimes, at 8 AM.

Vientiane's Thanon Thalat Sao or Morning Market is a sprawling marketplace just north of the city center, to the right off Lan Xang Road. Today, much of the market has been moved into a series of huge pavilion-style buildings, and it's here that you'll find souvenir Soviet military watches and watch repairers. The main attraction is the enormous amount of Lao textiles which fill lit-

OPPOSITE and ABOVE: Vientiane temple artistry. OVERLEAF: Wat Xieng Khuang (Buddha Park) is Vientiane's "Disneyland" of Buddhist lore.

erally hundreds of stalls. Anyone even re-
motely interested in textiles will want to
spend hours here, looking and examining
the wares. To the left side are stalls selling
skeins of hand-finished and dyed silk, all
ready for working into the handlooms,
while the creamy colored natural silks are
ready for dyeing. The market is also a haunt
of tribal women who come to buy new silk
and bring in their handloomed textiles to
sell. Ready-made silk clothing, tableware,
fabrics, and gift items are also available. Hill-
tribe silver can also be found here, although

is on a motorbike, which you can rent for
about US$5 a day (24 hours) from most
hotels. It's a relaxing city, where you can
buy a bottle of good French wine at a very
reasonable price at the **Vinothèque la Cave**
at 345 Samsenthai Road, stock up on baguettes together with some French cheeses
and pâté at one of the burgeoning mini-
marts (try **Phimphone Market** ((21) 216 963
at 110 Samsenthai Road) and prepare a picnic or a day's jaunt into the countryside.
Visitors can quickly feel immersed in the
everyday life, yet still be excited at the pros-

now the antiques have gone and much of
the silver is newly made. However, beautifully crafted silver and gold belts are still
readily available.

Just cruising around the city will introduce you to its special contrast of colonial
French and traditional Buddhist architecture; its people, who have a friendly dig-
nity that's a shock after the more established
tourist spots in Asia; its lifeline, the Mekong
River, which is an adventure all its own if
you follow the riverfront through the pro-
gression of bars and hotels that hug it on
the western stretch toward the airport; and
its rice fields and farming communities
along the southern road to Tha Deua and
Buddha Park. And the best way to do this

pect of new sights and encounters each day
you set out.

It is easy to spend a day browsing the
souvenir and textile showrooms — a visit to
the morning market (**Thanon Thalat Sao**)
with its myriad stalls, or to the internationally famous salon of **Carol Cassidy** ((21)
212 123 (for appointments) is a good place
to start, or try the **Art of Silk** ((21) 214 308,
supported by the Laos Women's Coopera-
tive, UNICEF, and SIDA, which has an extensive collection of antique handlooms in
its museum above the showroom. **Kanchana**
((21) 213 467, in That Dam Street opposite
the Ekalath Métropole Hotel, has a marvelous
collection of antique textiles and beautiful
new ones. In Samsenthai, the **Laos Cotton**

showroom has heavy handloomed cottons, perfect for furnishings, while **Doris Jewelry** ((21) 218 821 in Laos Hotel Plaza has marvelous textiles, antiques and jewelry. Another shop not to miss is in the **Lan Xang Hotel** ((21) 313 223, owned by the knowledgeable Mr. Bounkhong Signavong, which also has an extensive collection, while the vegetarian restaurant **Just For Fun**, on Pang Kham Road opposite Lao Aviation, displays some antique pieces and the Thai owner Yuwadee Silapakit will take you to her factory to see the weaving process.

not to mention billiard rooms and a health center. It's all quite reminiscent of Jakarta's old Hotel Indonesia.

Vientiane's hotel scene has perked up considerably with the advent of a first class 233-room **Novotel Hotel** ((21) 213 570 FAX (21) 213 572, on the airport road at Unit 9 Samsenthai Road, with rooms from US$45 to US$450. While the location is a little further than walking distance from the main tourist drag, the hotel offers international standards, a 24-hour café, a gym, business center and one of the few discos in town al-

WHERE TO STAY

The **Lan Xang Hotel** ((21) 214 102 or (21) 214 104 FAX (21) 214 108, overlooking the Mekong on Quai Fa Ngum, is the travel deal of Laos — or at least is was at the time of writing. With competition from the big new hotels, Lan Xang has reduced their prices to US$25 per night, which for a full-facility hotel is very reasonable. The hotel enjoys an excellent location within walking distance of Vientiane's important restaurants and shopping areas. Typical of government-run hotels, the relaxed and sprawling old hotel has several restaurants, a uniformed staff — all the men wear suits — massive gardens and a half-Olympic-size pool in the grounds,

lowed to play Western music, although food reports have not been that complimentary.

Close to the town center is the glittering new **Laos Hotel Plaza** ((21) 218 800 or (21) 218 801 FAX (21) 218 808 (for reservations), 63 Samsenthai Road in the midst of the tourist belt. The hotel incorporates a small shopping plaza with some upmarket art and book shops within. The big, cool lounge is a very pleasant place to visit for a drink, even for guests staying at other hotels. Rooms cost from US$100 to US$400 a night. The smallish and attractive **Tai Pan Hotel** ((21) 216 906

OPPOSITE: French bread alfresco and shoppers at the Morning Market. ABOVE: The Asia Pavilion Hotel (formerly the Constellation) in downtown Vientiane.

FAX (21) 216 223, 2-12 François Nginn Street, costs from US$50 to US$70. The slightly seedy exterior belies the comfortable interior of the **Asia Pavilion Hotel** ((21) 213 430-1 FAX (21) 213 432, centrally located at 379 Samsenthai Road, which has been quite expensively renovated since its pre-revolution days as the French Constellation Hotel, mentioned in John Le Carre's novel *The Honorable Schoolboy*. It caters to a lot of tour groups. Most rooms have a refrigerator and television. Rates range from US$18 to US$35.

Le Parasol Blanc ((21) 215 090 or (21) 216 091 is a quiet and comfortable garden hotel, with chalet-style rooms built around an old colonial villa set beside a small swimming pool. Located on Nahaidio Street, just past the Pratuxai Monument, Le Parasol is popular with consular and United Nations personnel, and charges US$33 a night for a double room. The garden restaurant serves good Lao and pleasant western-style food in a very comfortable setting.

There are two quite deluxe guesthouses in Vientiane, both of them restored villas in garden settings. **Lani One Guesthouse** ((21) 216 103 FAX (21) 215 639, 228 Setthathirat Road, next to Wat Hay Sok, has 11 rooms, and you'll find **Lani Two** ((21) 213 022 FAX (21) 216 095 with seven rooms at 268 Thanom Saylom. Both charge about US$30 a night for a double. The **Royal Hotel** ((21) 214 455 FAX (21) 214 454 on Lan Xang Road has nice rooms in a glittering Asian kind of way, with rates from US$80 to US$130.

WHERE TO EAT

One of the marvelous things about dining in Vientiane is not just the abundance of delicious Lao and Vietnamese food available, but also the number of restaurants serving authentic French and Italian dishes, so that visitors are really spoiled for choice. The alfresco garden restaurant and piano bar at **Le Parasol Blanc** (see WHERE TO STAY, above) is a pleasant and relaxing experience with large-sized rattan chairs and where food ranges from baguette sandwiches to Lao specialties to very good, authentic pizzas. French and Italian restaurants worth a try include the **Nam Phu Restaurant-Bar**, ((21) 216 248 in Nam Phou Circle opposite the fountain. The

specialty is French cuisine, although they will provide Lao food on demand, and the service is excellent. It is very popular with visiting French tourists and diplomatic officials, along with resident expatriates. Right opposite Nam Phu, the very popular **L'Opera** ((21) 215 099, an Italian restaurant with an Italian chef, is reputed to be the best Italian in town, with good espresso coffee and a wide selection of Italian wines. A newer Italian addition is **Lo Stivale** (215 561 at 44 Setthathirat Road, which serves great pizzas and pasta. Back to Nam Phu Circle is the **Restaurant Le Provençal** ((21) 217 251 with southern French provincial fare and a good bar. Also worth a visit is **Le Bistrot** ((21) 215 972, on François Nginn Street opposite the Tai Pan Hotel, for *cuisine familiale* at very reasonable prices, along with popular French dishes and couscous. A quite trendy restaurant is the somewhat quirky **Le Vendôme** ((21) 216 402 in the tiny Wat Im Paeng Lane behind Wat Im Paeng where the food is quite good and the candlelit atmosphere quite interesting. Other French and Western establishments are the old timer **Arawan** ((21) 215 373 at 478 Samsenthai Road, which also has a well-stocked *charcuterie* and specializes in French favorites like *coq au vin*, and French wine and cheeses. The **Souriya** ((21) 215 887 at 31 Pang Kham Road has a reasonable and quite pricey French menu.

The rather tacky **Fountain Bar** set around the tired looking central fountain in Nam Phou Place, offers open-air dining with Lao and Chinese food, along with a very welcome cooling spray on hot nights.

For Lao and Chinese food, there are a great many places to choose from and no real need to list them. All the hotels do great Lao food for starters, as does the backpacker **Patuxai Café** on the Mekong — particularly popular at sunset. Quality baguettes are available all over town and with a choice of fillings, although personally, I would prefer to buy the fillings from a mini-mart. Just north of the Patuxai Monument, the popular **Laos Residence** is housed in a beautiful garden villa and is very popular with tour groups, although the food is modified to suit tourist tastes. Le Parasol Blanc does excellent Lao food, otherwise I prefer to try what comes along.

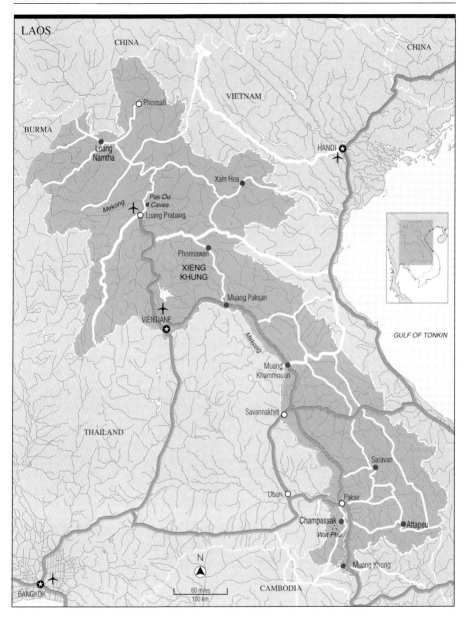

The **Noorjahan** at 370 Samsenthai Road and **The Taj** on Pang Kham Road, just north of the fountain square, are excellent Indian restaurants. **Just For Fun** ((21) 213 642, opposite Lao Aviation on 51 Pang Kham Road, is just for vegetarians with a range of herbal teas and meatless dishes.

Street food is also worth trying but just make sure that the vendor and her shop look clean. The food is usually very cheap, very authentic and generally delicious, with the

added benefit of offering an occasion to meet some very nice Lao people.

Samsenthai has several restaurants and one hole-in-the-wall establishment opposite the Asian Pavilion Hotel serves up excellent Vietnamese food, not the place for a fancy meal, but certainly the place for a satisfying dinner or lunch. One of the best meals I had was outside the morning market. After buying a fresh baguette, I followed my nose to a place barbecuing fragrant marinated pork

and indicated that I would like to try it, much
to the amusement of the Lao onlookers. The
pork was barbecued and brought over with
some fresh Vietnamese-style salad and a cup
of thick Lao coffee — it was hard to beat.

HOW TO GET THERE

Although the international Wattay Airport
was closed to all except Thai International
and Lao Aviation flights from Bangkok for
much of 1998 in order to upgrade the runway and terminals, for Visit Lao Year 1999,
Vientiane will again be accessible from Singapore by Silk Air and Malaysian Air Services
which flies via Kuala Lumpur and Phnom
Penh. From Phnom Penh flights are by Royal
Air Camboge and Lao Aviation, and from
Saigon and Hanoi by Vietnam Airlines and
Lao Aviation in a joint-venture deal. Thai
International and Lao Aviation operate daily
flights to Vientiane from Bangkok.

Entry to Laos is also possible by traveling by overnight train from Bangkok to Nong
Khai, then driving across the Friendship
Bridge into Laos, collecting a visa on arrival.
If you're on a group or individual prepaid
tour, you'll find a local tour agency mini-van
or car waiting to take you the 20 km (12 miles)
to Vientiane. If not, taxis are available which
charge about US$5 to take you to town.

Lao Aviation operates all internal flights
in Laos with a mixture of Russian and Chinese
prop-driven aircraft, and some newer aircraft. Since Laos has opened up, it is possible
to book reasonably-priced flights all over the
country.

ENVIRONS

The rather bizarre Buddha Park is actually
called **Wat Xieng Khwan**, but bears special
mention because it's not a temple at all but
rather a kind of showcase of old-style Lao
Buddhist and Hindu imagery set in a pleasant tree-shaded park on the banks of the
Mekong, south of Vientiane beyond the Tha
Deua border post. Crowded with just about
every sacred image from the two religions
that you can name, and dominated by a huge
reclining Buddha, it was designed and built
in 1958 by a mystic named Luang Pu who
founded a cult based on a mixture of Bud-

dhist and Hindu theology. After the 1975
revolution, he moved across the river to continue his teachings in Nong Khai. The Buddha Park is fascinating as a photo-opportu-
nity and a pleasant place to spend an hour
or two. Small stalls provide simple refresh-
ments. You can take a car or *tuk tuk* for the
24-km (15-mile) ride, or go for the real adventure and take the local bus from Thalat
Sao Bus Terminal.

LUANG PRABANG

To arrive in Luang Prabang by plane is to fly
into one of Asia's most beautiful cities — a
big tropical garden in the midst of a moun-
tain-enclosed valley. It is to see the waters of
the Mekong and Nam Khan rivers flashing

amidst the palms and other foliage, old French villas and civic buildings nestling here and there in the trees, and the gold-leafed spires of its many temples sparkling in the sun. For many years, Luang Prabang ranked with mythical Shangri-La as a fabled but virtually forbidden destination in Asia — protected by ramparts to the south, it was inaccessible by land and difficult to access by river. Even during the war it was difficult, and out of reach completely during the revolutionary years. Only now is it wide open to foreigners to the point where there is a new and potentially disastrous (from a culturally invasive point of view) daily flight from Bangkok on Bangkok Airways.

From the ground Luang Prabang is even more idyllic with a charm that is almost palpable. The road into town from the airstrip passes rows of neat timber shop houses, temple walls and weathered colonial buildings, all set among trees, with the city's highest pagoda, That Chom Si, atop a steep hill in the center of town, providing a dramatic backdrop all the way. To describe this royal town as a laid-back place would be an understatement: the people look as though they're enjoying a permanent siesta. The only real activity occurs in the early morning when barefoot monks roam the misty streets, their saffron robes adding color to the gray dawn, treading softly, in long orderly lines on their daily alms rounds.

Situated on a peninsula between the junction of the Mekong and smaller Nam Khan

Luang Prabang — "one big tropical garden."

Rivers, Luang Prabang consists of streets of intact colonial architecture standing side-by-side with ancient wats.

In February 1998, it celebrated the official consecration as a UNESCO World Heritage Site, which had been conferred in 1995, adding to an impressive list that already included Angkor Wat, Quebec City, Jerusalem and India's Taj Mahal. They cited Luang Prabang as the best preserved historical town in Southeast Asia, recognizing its great historical and cultural significance with 33 wats and 111 French-Lao buildings cited for specific conservation. This kind of recognition will help protect the town from the depredations of too much tourism and even now, no out-of-character buildings are allowed, no advertising billboards, no big tourist hotels within the boundaries of the old city. Power and telecommunication lines are to be buried out of sight. It also means that many more of the ancient wats will be restored.

The downtown "business" district, where you'll find Lao Aviation, Inter-Laos Tourism, the General Post Office, the National Museum and the morning market, lies beyond the oldest quarter and is not so protected.

Converted villas like the Villa Santi and Le CaLao are situated right in the heart of the old town, making a convenient base to explore the most interesting sites.

You can walk or bicycle all over Luang Prabang, quickly becoming familiar with its simple grid of streets and main attractions. The people are friendly but also quite reserved: they nonchalantly accept foreigners in their midst, without getting excited, which makes it very pleasant to move around.

As this is the Angkor Wat of Laos, it's very much part of a daily tourist conveyer belt that runs from Vientiane, with Lao Aviation operating daily flights full of tour groups and even pulling aircraft off other routes to provide extra services if the traffic gets too heavy.

In spite of being a popular destination, it is far from overcrowded and the charm of this royal town seeps into your consciousness the longer you stay. The slow pace, the evening chanting of the monks and the dawn alms parade through the cool streets provide an enchanting backdrop that will make you never want to leave. While organized tours to Luang Prabang are readily available, this is one town where a tour is not only unnecessary, but could be downright irritating. The town is made for a slow pace and ultimately to be enjoyed with the same degree of grace that the town exudes.

With such a wealth of fine temples and buildings you can get quite dizzy trying to cover everything in a hurry. Better to take time, and explore at your leisure, those buildings that hold special appeal.

WHAT TO SEE AND DO

The **National Museum** makes a good place to visit first. This former royal palace, is an opulent French-Lao mansion backing on to the Mekong. Originally constructed in 1904 as the official residence of King Sisavang Vong, father of the ill-fated last monarch, it is literally packed with precious Buddha statues of all descriptions and treasures from the dynastic era. The main hall, with its gold-painted walls, handloomed curtains from India and obvious opulence are only topped by the blood red walls and mirrored decorations of the receiving room where gasps of awe can be heard at the entrance by impressed visitors.

Some of the Buddha images appear to be priceless, including a reclining Buddha and a standing image made of marble. Other images and treasures fill the large reception halls to the right and left of the entrance. To the right, the king's reception hall features busts of the various Lao monarchs, screens depicting the *Ramayana* and walls decorated with dramatic murals of traditional Lao life painted 70 years ago by the French artist Alex de Fautereau. Beyond that, in another exhibition room, a large gold Buddha called the Pra Bang (from where Luang Prabang takes its name), presented to Fa Ngum by his Khmer benefactors when he conquered Luang Prabang, stands amid ivory Buddhas fashioned from elephant tusks, a host of other sculptures, a temple frieze and three embroidered silk screens featuring Buddhist stories.

To the right of the entrance, next to the queen's reception hall, is a room crowded with exhibits of various official gifts given

The beautiful National Museum, formerly the Royal Palace.

by foreign heads of state and VIPs over the years. What's fascinating about it is the sort of kitsch that royalty and government leaders give each other — stuff that most people would probably keep in the attic or unload at a garage sale. Amongst all the wealth and finery, there is even a scale model of the Apollo moon vehicle *Columbia*, presented by Richard Nixon.

Wandering through the royal apartments to the rear of the throne and reception rooms, the human element appears in old sepia photographs of the royal family in happier

The markets are always a focal point in smaller towns or rural areas. In Luang Prabang, they are populated by tribal folk coming in from the villages to sell their wares and produce. You will always see a few interestingly-dressed folk around the steps of the main market, the **Thalat Dala**, in the center of town, whose dim interior is filled with stalls selling textiles, hardware for use by the farming communities, silver jewelry and souvenirs, watches and a host of cookware and basic foodstuffs. The bustling morning market, **Thalat Sao**, at Wat Pasaman Road near Wat

times, especially a misty Dietrich-style print of the queen. It counterbalances the moral and political issue of a royal entity who gave his royal blessing to the United States bombing and counter insurgency operations in Laos during the war.

After all the finery of the receiving rooms, the bedrooms at the rear are surprisingly austere, devoid of any personal touches at all and much less luxurious than the average hotel room. No one outside the highest echelons of the government really knows what became of King Vatthana and his queen, but it's quietly whispered that after their exile to the Sam Neua Province, they were later to be put to death in 1977 for refusing to support the revolutionary regime's policies.

Phraphouthabat, is perhaps the most colorful and is far less visited by tourists. This is strictly a produce market supplying most of the town's needs. A small but busy market fills the street leading up from the river to the Phou Si Hotel, filled with river folk each with their piles of fresh produce to sell and vendors selling bowls of steaming *pho* noodles.

Luang Prabang has so many temples that to visit the town and not see at least a few wats would seem to be somewhat churlish. While one could spend a week or two visiting them all, several stand out as particularly interesting.

The commanding view of the town from **That Chom Si**, on its high perch atop the peak of Phu Xi (or Phou Si) is worth the climb while the That itself is relatively new, built in 1804. The panorama from the lower terrace of its stupa is wonderful at sunset, imparting a sense of serenity as the colors change right across the city and the river stretches. Standing on a ridge close to the temple is an old revolving Russian antiaircraft gun — a leftover from the early revolutionary days.

To the north of the peninsula, **Wat Xieng Thong (Golden City Temple)** is located right on the confluence of the two rivers, and is considered to be the city's finest. The wat is best visited late in the afternoon, when the golden light shows off the temple to best advantage, the gilded carved doors glowing in the late afternoon sun. The large complex of shrines, pavilions and prayer halls feature, amongst other things, brilliant and quite unusual mosaics and rare, gilded erotic carvings illustrating excerpts from the *Ramayana*. Among the other treasures at this 400-year-old site are a pavilion packed with Buddha images and a huge, gilded royal funeral chariot with dragon heads rearing from its prow, another containing a reclining Buddha dating back to the temple's construction in 1560 and, at one corner of the complex, a royal barge.

The oldest temple site, **Wat Visoun**, located south of Wat Phou Xi near the Rama Hotel, dates back to 1513, shortly before the first Lao kingdom of Lan Xang was established here by the warlord Fa Ngum. The city's glory lasted only 12 years, until 1545 when the capital moved to Vientiane, but it remained a seat and power base of Lao royalty right up until 1975 when the Pathet Lao hauled the last monarch, Savang Vathana, off to probable execution. Its main features are a collection of fifteenth-century ordination stones, a display of wooden Buddhas sculpted in the "Calling for Rain" posture. The dramatic white **That Pathum** (Lotus Stupa) in front of the main hall, is known more popularly as **That Mak Mo (Watermelon Stupa)** because of its bulbous shape. The collection of small gold and crystal Buddhas in the throne room of the National Museum is said to have been found in this stupa.

Wat Mai, close to the General Post Office off Luang Prabang's east-west main street, Phothisarat Street, is a relatively new temple, built in 1796. Its five-tiered roof and gilded,

sculptured door panels depicting scenes from the life of Buddha and the *Ramayana* are among its key architectural features. The compound also houses two shallow-draft traditional barges which lead celebrations on the rivers during the Lao New Year in April and Water Festival in October. **Wat That Luang**, built in 1818 on a hill to the east of the city, has a central stupa containing the ashes of King Sisavang Vong.

Across the Mekong River from central Luang Prabang you'll find a complex of smaller temples that include **Wat Tham**, built into a limestone cave, and **Wat Chom Phet**, which provides another dramatic panorama of the town and river. Take a small boat from the bottom of the steps at Wat Xieng Thong.

WHERE TO STAY

Probably the best address in Luang Prabang is **Villa Santi (** (71) 212 267, previously known as Villa de la Princesse, set in a stylishly renovated former royal mansion on Sakkarine Street, one of the gems of Luang Prabang's tourist industry. It has rooms that are deluxe by Lao standards, furnishings and decorations that are said to have come from the royal collection, a quiet grassy courtyard for cocktails and buffets, an upstairs verandah bar and dining room with a view of the surrounding streetlife and regular performances of traditional Lao court and tribal dancing. The rate is US$40 a night but it is necessary to book well in advance as it is almost always fully booked.

Elsewhere, the sprawling **Phou Si Hotel (** (71) 212 192 or (71) 212 717 FAX (71) 212 719 (or in Vientiane **(** (21) 213 633/4) enjoys a convenient location at the town's main intersection where Phothisarat and Setthathirat streets meet. This pleasant hotel is set in tropical gardens reminiscent of Bali with rooms in the US$28 to US$56 range and an open-air bar. The **Phou Vao Hotel (** (71) 212 194 FAX (71) 212 534, set on a hill to the east of the town, has a swimming pool and deluxe rooms at between US$25 and US$60 a night. It also has superb views of Luang Prabang.

OPPOSITE: Meo hill tribe women LEFT in Luang Prabang. Traditional weavings RIGHT have become a tourist attraction at Ban Phanom, outside Luang Prabang.

Another converted villa guest house is the elegant and tiny **Le CaLao Hotel** (/FAX (71) 212 100, close to Wat Xieng Thong, with just five spacious rooms. Ask for one of the four rooms upstairs US$40 that overlook the Mekong. Advance booking is required.

Yet another charming hotel, but located at the other end of town, is the 24-room **Hotel Souvannaphoum** ((71) 212 200 FAX (71) 212 577, in Phothisarat Street, belonging to a different member of the royal family which has rooms for US$55 per night. Rooms are centered around a sprawling garden and the

big, charmless rooms for US$10 a night which are fine in an emergency (i.e. when all the better hotels are full). It has Luang Prabang's only (very noisy) discotheque and provides an indelible cultural experience as young Laotians get it on, together with the odd Western visitor.

WHERE TO EAT

As Luang Prabang blossoms into the most attractive destination in Indochina, the restaurant scene too, is blossoming. The pick of

cuisine is Lao and French. The French-run and owned **Duang Champa Auberge and Restaurant** (/FAX (71) 212 420 lies in a converted villa overlooking the Kahn River. The spacious rooms cost a very reasonable US$20 per night. A newer option is the 35-room **Muong Luang Hotel** ((71) 212 790/1 FAX (71) 212 790, designed in traditional Lao style; the restaurant serves traditional Luang Prabang and Lao dishes. It is located at Bounkhong Road and is one of two hotels with a swimming pool. The family-run **Vanvisa Guesthouse and Antique Shop** (/FAX (71) 212 500 is in another old converted villa in Ban Watthat 42/2 with rooms for around US$35. The basic **Rama Hotel** ((71) 212 247 has the advantage of being right in town and offers

the best includes the French **Auberge Duang Champa** ((71) 212 420 overlooking the Nam Khan River. Located in an old French Villa, it is run by a Frenchman and his charming Lao wife, who offer a good French menu, starting at amazingly low prices of around US$2 for a set meal (à la carte costs more) and an acceptable wine list. The appealing **Nam Kharn Garden Restaurant** overlooking the Nam Khan River offers Lao and Western dishes, French wines and a pleasant view of the river. Particularly interesting is their *khai paen*, a fried seaweed wafer

OPPOSITE: Prayer hall in Wat Xieng Thong. ABOVE: Poolside LEFT at the Luang Prabang Hotel. Five-star comfort RIGHT of the Villa Santi Hotel.

and a Luang Prabang specialty. The elegant dining room at **Villa Santi (** (71) 212 267 is a popular place to eat, with authentic Lao and French cuisine cooked by the daughter of the chef to the last king. **Le Saladier** across from the Rama Hotel offers an elegant menu, the nicest of a strip of restaurants that run along this street.

ENVIRONS

The Buddhist grottoes of **Pak Ou Caves** lie 30 km (19 miles) west along the Mekong River, and the trip there provides another insight into the beauty of this great waterway, especially in the dry season when the skilled boatmen effortlessly dodge currents and whirlpools created by the low water. On the way, the squat, shallow-draft tour boats call into a small village which is famous for its traditional rice wine stills. You can watch the fiery concoction being distilled and also enjoy a tipple or two, for- tifying yourself with the powerful essence for the rest of the trip ahead. Right in the midst of a sheer limestone cliff are the two Pak Ou Caves — an astonishing sight. A short climb brings you to the main cave and a sight that burns into the memory banks. Hundreds and hundreds of **Buddha images** of all shapes, styles and sizes stand on ter- races within the cave, held under the scrutiny of a local man who stands guard. There used to be a lot more statues, but Pak Ou was one of the prime targets of thievery and smuggling rackets which saw many of the treasures of Laos spirited out to the United States and the West during the war.

Beyond Pak Ou there's another spectacu- lar limestone cliff where the **Nam Ou River** joins the Mekong, and tour boats sometimes make a short trip along there before return- ing to Luang Prabang.

The small village of **Ban Phanom**, just a few kilometers east of the city center, is Luang Prabang's favorite souvenir stop, where all tours, individual or group, inevi- tably end up, and it is well worth a visit.

Although the clacking of handlooms has been more or less replaced by the quiet sounds of money changing hands as tour- ists make their purchases, the quality of the handloomed pieces is high, while the prices are low. Ban Thanom is populated by Lu minority people, who were once responsible for weaving the fine textiles required by the royal family. Today it is famous for its qual- ity **cotton and silk weaving**, and textiles are imported from other parts of the country to satisfy the ever growing market. You can still find the odd woman working on

traditional looms, while others are busy haggling with the buyers in the busy mar- ket place that has sprung up in the center of the village. The village is part of a very sensible local government campaign to re- store traditional arts and crafts. To get there take a jumbo from the main market, rent a bicycle or just flag down a passing *tuk tuk.* It should cost just a few dollars to get there and back.

A pleasant drive 29 km (18 miles) south of Luang Prabang brings you to the beautiful **Kuang Xi Falls**; the drive makes a delight- ful excursion from the city. The lower falls have become a popular spot for picnics while a second, less visited fall is accessible by a small trail.

PREVIOUS PAGES: Gilded artistry of Wat Sen, typical of the Buddhist heritage of Luang Prabang. OPPOSITE: Buddha images fill the sacred Pak Ou Caves down river from Luang Prabang. ABOVE: A Luang Prabang village hut.

HOW TO GET THERE

Most visitors take a flight from Vientiane or even from Bangkok by Bangkok Airways. Luang Prabang makes a triangle with Oudomxai and the Plain of Jars so it is quite convenient to fly there before or after Luang Prabang. An increasingly popular route is to enter Laos from Huay Xai in the very north and take the Mekong to Luang Prabang. Many tours are available for this route. From Vientiane, the road has been officially de-

LUANG NAM THA

A flight slightly over an hour long from Luang Prabang over rugged mountain ranges brings you to the small airport of Luang Nam Tha, capital of Nam Tha Province, bordered by Myanmar to the northwest and China to the north. The province is home to 39 ethnic minority groups, who inhabit the remote mountain areas. While the town appears quite attractive, it is the villages out of town, and particularly the old trading town of

clared safe for travel, and it is now reputedly possible to drive from Vientiane.

NORTH OF LUANG PRABANG

As the remote north of Laos opens up to tourism, the current basic facilities will no doubt improve. This is adventure territory — a land of rugged, jungle-covered mountains peopled by hill tribes living in small villages. Getting around requires public transportation, a motorbike or the services of a tour company, which is probably the best way to visit. A tour company like Diethelm or Intrepid will certainly maximize your time and ensure that you get to visit interesting villages (see TAKING A TOUR, page 67).

Muang Sing that are really worth the visit. That said, Luang Nam Tha has plenty of reasonable accommodation.

MUANG SING

Newly discovered by the tourist trail, the old trading settlement of Muang Sing that dates back to the sixteenth century is preparing to become a shining light in the northern Laos tourism stakes. Known for the numerous tribal folk who inhabit the villages surrounding the broad rice plains of the Nam La River, Muang Sing was once the major opium market of the northern highlands, and is located just a few miles from the Chinese border. The main attraction is the colorful daily market attended by a mixed crowd of tourists and T'ai Lu, Zao, Iko, Shan, Lao Sung, Hmong and T'ai Daeng folk who converge at the market for trade and pleasure, and by ten in the morning its all over. Women set up stalls in the cold morning light doling out steaming bowls of turkey pho—a delight both for the villagers and visitors. Other women sell fried donuts or approach tourists to sell their hand-embroidered head wraps at bargain prices, and a great time is had by all. Accommodation in Muang Sing is basic and costs around US$1 to US$2 a night, but a little discomfort is worth it for the visual adventure. No doubt, as the tiny market town becomes better known, more upmarket rooms will be available. Outside the town it's possible to visit tribal villages, often a few kilometers' walk from the main road.

The best time of the year to visit is during the **Muang Sing Festival** held on the full moon of the twelfth lunar month, which occurs around late October or early November, depending on the vagaries of the moon. During the festival, villagers come to town resplendent in their best costumes, for several days of celebrations. This Buddhist-animist festival centers around the wat, on a sacred hill to the south of town, starting a few days ahead of the full moon. Traditional dances, games and Laotian pop music entertain.

OUDOMXAI

This burgeoning capital of Oudomxai Province that sits between Luang Prabang and Luang Nam Tha also shares a border with China's Yunnan Province and is home to 23 ethnic minorities including the Akha (Iko), T'ai Lu, T'ai Dam, T'ai Neua. Oudomxai town has a major Chinese influence, in part because imported skilled Chinese labor is involved in a major road building projects to make it the northern commercial and trading hub, linking the town to surrounding provincial centers. As such, the brash town lacks the graciousness of other centers and many of the hotels function as brothels for the Chinese workers. However, the market is good and the town is well located for trekking and hiking through remote mountain villages. It is possible that as canny tour

Sunset on the Mekong.

operators become aware of the potential, conditions will improve drastically. Oudomxai has flights to Luang Prabang and Vientiane from a newly-built airstrip.

XIENG KHUANG (PLAIN OF JARS)

One of Laos's more enigmatic destinations, the desolate **Plain of Jars** has caught the imagination of visitors since it was opened to tourism in the early 1990s. **Phonsavan**, the new capital of northern Xieng Khuang Province, reminds me of the rugged, undeveloped frontier towns that you find on the lower Tibetan Plateau in China. The city was built after the former capital, Xieng Khuang was virtually destroyed by bombing during the campaign against the Communist Pathet Lao in the Vietnam War. It's a sprawling and flat nondescript market town of cement and corrugated iron — a study in charmlessness which seems to be a Communist specialty. The town lies at the center of a vast, defoliated, utterly devastated, dusty plain that still shows scars of some of the heaviest bombing of the war. In the dry season the winds that sweep across the area are distinctly cold in the daytime and bitter at night.

Phonsawan is the gateway to the country's unique historical attraction, the large stone jars that lie scattered across these dry plains. Weighing as much as six tons, these mysterious vessels point to the sky like fat siege mortars, which more than one observer has cynically suggested the United States bomber pilots mistook them for. The ground has been littered with UXO (unexploded ordnance) and has been extremely dangerous especially to children playing in the fields. Although it is being slowly cleared, visitors are cautioned to keep to the marked paths. While in Phonsavan, though, we met with British UXO experts who were celebrating the end of their assignment and had cleared large tracts of the land.

The jars are said to be many hundreds if not thousands of years old, but beyond that no one has really come up with the definite explanation for them, although theories of their origins abound. One suggests that they were wine fermentation jars put there by a sixth-century resistance hero to celebrate his victory over a despotic local ruler. Another suggests they were burial jars. Yet another explanation may well lie in the region's arid character during the winter months — were they, in fact, nothing more than water storage vessels? Whatever the answer, they tease the imagination, and they are one of those strange cultural attractions which draw visitors simply to be able to say they've seen it.

According to local reports, there were once several hundred jars at several sites across the region, but the smaller ones have been souvenired and others shattered so that less than 100 now remain in main site, in two groups scattered on a hillock and across the floor of a shallow valley. This main site **Thong Hai Vin** or Stone Jar Plain is 12 km (seven miles) from town.

Scattered across the hilly plain were about 250 jars ranging in size from 600 kg (1,300 lbs) to the largest which is almost six tons, all carved from solid stone. While the smaller ones have been souvenired, about 100 remain, mystifying those that gaze upon them. Site Two is 23 km (14 miles) from town where smaller jars are scattered across two hillsides, while Site Three is 28 km (17 miles) from town and accessible by a delightful walk past a small wat and about a kilometer through verdant rice fields if it is the wet season. Here strewn across a grassy hillock is the most picturesque site although the jars are smaller and less dramatic than the main site. Other smaller sites are hidden in the hills but be warned — not all have been cleared of ordnance and could be dangerous.

The flight from Vientiane is certainly something not to be missed — carrying you over the vivid physical contrasts of the rice-plains around Vientiane, then incredibly beautiful mountain ranges featuring many rivers and literally dozens of potential white-water rafting spots, and a vast reservoir, Ang Nam Ngum, and hydroelectric plant which not only serves domestic power needs but exports electricity to Thailand. Quite frankly though, Phonsawan itself has little to offer aside from spectacular landscapes of lush rice terraces, especially in October and November after the wet season, and the mysterious novelty of the Plain of Jars. Most visitors find

OPPOSITE: The fabled Plain of Jars TOP. Market scene in Xieng Khuang BOTTOM.

that one night is sufficient to explore the Plains with time enough to make it out to the old ruined capital of Xieng Khuang where a lonely bombed Buddha surveys the surrounding land.

Those not on an organized tour will find getting around to be quite easy. Local tour guides meet every flight at the airport, drumming up business and will take you on a hotel tour until you find one to your liking and then provide transport to the various sites. Jeeps are available and minibuses which cost US$55 for the trip to the plains—significantly cheaper than a tour. On my journey, all eight very disparate passengers made up an impromptu and very amenable group that lasted for the length of our tour. Our guides took us to several hotels until we found one which met with everyone's approval with budget rooms as well as five comfortable "VIP" rooms at the rear which came complete with hot water for an hour or two.

WHERE TO STAY AND EAT

Phonsavan's tourist facilities are basic but with a little effort, it is possible to be comfortable. There's a modern 50-room hotel called the **Plain of Jars Auberge**, run by Sodetour, but since the French owner was tragically killed by insurgents, it is rumored that the place has gone downhill. Used by tour groups and located near the jars themselves, at last report they were charging US$40 to US$50 a night for double rooms (no phone number available). The reportedly rundown **government guesthouse** on the western edge of town stands alone on a hilltop in a bleak landscape that reminded me of outback Australia without the eucalypts. It's closed a lot of the time and only opens when tourists turn up, and its rooms are basic with a communal cold-water bathroom. Electricity is available only from 6 PM to 9 PM. But apart from that, it's not such a bad place, and it's a good viewpoint for savagely beautiful sunsets created by the region's dust. We stayed at the large Communist-inspired **Phou Doi Travel Hotel** ((61) 312048 in Nam Kath Road with five comfortable rooms for US$5 to US$10. The **Vinh Thong Guest House** on Road No. 7 also looked quite comfortable.

In "downtown" Phonsavan, near the very interesting market if there is time to visit, the **Hayhin Hotel** charges in kip and a room will cost about US$2 a night. Right across the road, the **Phonesaysouron Restaurant**, serves a fixed but wholesome Lao supper and breakfasts of thick coffee and fried sesame buns. We ate at the **Sangah Restaurant**, which offered simple but delicious Chinese, Lao and Thai-style dishes for just a few dollars.

PAKSE AND THE SOUTH

Pakse is the relatively prosperous administrative center of the Champassak Province and a lively market town. Lying in a balmy, subtropical landscape of rice paddies and groves of palms at the confluence of the Mekong and Se Don rivers in southern Laos, it was built as recently as 1905 as the new capital of Champassak Province. While it has very little history of its own, it is the gateway to the former royal capital of Champassak and one of the most important religious sites in Southeast Asia — the pre-Khmer Hindu/Buddhist temple of Wat Phu. Pakse is also a key access to one of the widest and most picturesque sections of the Mekong River and the 4,000 islands close to the Cambodian border. Until February 1998, Pakse market was a popular stop off point for textile collectors who came in search of the hand woven silk *ikats* from the surrounding villages. Unfortunately the market burned to the ground leaving a lot of unhappy and poorer (although still alive) people in its wake.

It is possible to enter Laos from Ubon Ratchahani in Thailand, crossing by ferry at Chongmek, across the Mekong from Pakse thereby covering Pakse, Wat Phu and the southern islands before heading north to That Lo and on to Vientiane.

WHAT TO SEE AND DO

Pakse itself is not a major sightseeing area but it makes a convenient base for exploring the surrounding areas. To the north is the **Boloven Plateau** home of tribal villagers, elephant workers, coffee plantations, the market town of Se Kong and the Se Kong River, waterfalls and the pleasant That Lo

Resort. To the south is **Champassak** and the pre-Angkor **Wat Phu**, the 4,000 Mekong Islands or **Si Phan Don**, and the remote and charming little town of **Attapeu** which is close to the Ho Chi Minh Trail.

Pakse itself is the last post of civilization for the south with some reasonable restaurants and hotels, a big and very lively market which shouldn't be missed and an interesting new museum with an ethnological section exhibiting the tribal peculiarities of the southern minority people. Easy half-day trips can be made to the worthwhile weaving village

of **Ban Saphay** 16 km (10 miles) north of town on the Mekong, where handloom silk is a specialty You can rent a jumbo to get there. The **Taat Sae Waterfall** is off Road No. 13, south of Pakse.

CHAMPASSAK AND WAT PHU

The docking point for the journey to Wat Phu, **Champassak** has little to show of its former glory as a royal capital, but archeological teams are busy exploring the former city's ramparts. This sleepy little ghost town exhibits some striking examples of old French civic architecture which stand virtually cheek-to-jowl with traditional and modern stilted, tin-roofed, wooden Lao homes. The

town lies on either side of a red-dirt road that runs parallel with the river until it makes an abrupt right turn and becomes part of the triumphal approach to Wat Phu. Along the way you may see two gaudily decorated Buddha images almost hidden by the encroaching branches of Banyan trees. No doubt there is a story attached to them. In the wide plain that lies before Wat Phu, another decrepit royal villa — a summer pavilion — lies on the edge of a large lotus pool, left to fall into ruin since the last king of Laos, Savang Vatthana, and his queen disappeared into the murky gulag of socialist reeducation.

The ruins of the ancient pre-Khmer temple **Wat Phu** lie 37 km (23 miles) from Pakse just past the town of Champassak — once the capital of the Champassak Kingdom and earlier, part of the Khmer Angkor Empire. Situated on the lower eastern slope of the sacred mountain of Phu Kao, whose peak is said to resemble a linga, it has been a holy site for millennia. The mountain is essential to Wat Phu's special status as a religious site — the monolith on top of it attracting Khmer Shivaites who built the first temple there well before the rise of the Khmer Empire. In the sixth century, a Chinese chronicle spoke of a temple on the site guarded by a thousand soldiers and dedicated to a spirit to whom the king offered a human sacrifice each year. Evidence suggests that it was also the principal temple of the capital of Chenla, Shreshthapura, which is believed to have been located on the site of present-day Champassak. As its name suggests, this in turn may previously have been part of the central Vietnam kingdom of Champa.

As possibly the oldest religious site used by the Khmer Hindus, Wat Phu is significant indeed. And its antiquity is all the more pronounced by the state it's in today. Of its two imposing main palaces, built on a terrace at the foot of the hill, only the outer walls are still standing and their entrances, featuring elaborately sculptured gables, have all but collapsed — the huge stones tumbling as though struck by an earthquake. Beyond them, a series of steep stone stairways, also in a state of ruin, lead up the hillside to ter-

Pakse woman prepares sticky rice.

races where pavilions, a library and as many as six other buildings once stood. The main sanctuary, located on the highest terrace, is relatively well preserved, featuring an ante-chamber and side-naves with walls and lintels decorated with carved *devatas* and *dvarapalas*. Close by, against the foot of a cliff, there's a bas-relief carving of Shiva flanked by Brahma and Vishnu and, nearby, a huge flat stone with the outline of a crocodile carved deeply into it.

Guides will enthusiastically demonstrate how the outline neatly embraces the human

a popular commercial venture in recent years with local teenagers wandering about and loud music that manages to detract from the holy atmosphere one may be expecting.

Another festival held each June climaxes with the sacrificial slaughter of a buffalo.

While Wat Phu is in less than perfect condition, it is well worth a visit for the atmosphere and the sense of history that pervades it. Nearby is a pavilion erected by a team of United Nations working archeologists who are resurrecting the surrounding statuary.

body, and how a channel in the stone was put there to drain the blood — supporting the theory that this was in fact the altar upon which the early monarchs performed their human sacrifices.

The present ruins of Wat Phu are said to date back to the eleventh and twelfth centuries. Three Buddha images now stand in the sanctuary, and it is suggested that this may well be Southeast Asia's oldest Buddhist temple. But throughout the complex, much of the architecture and surviving decoration is definitely Hindu.

The temple's key festival, staged in the three days leading up to the February full moon, coincides with the Buddhist Makha Puja, although unfortunately it has become

The best way to get to Wat Phu is the languid trip by boat along the Mekong from Pakse, and this trip, favored by the tour agencies, usually combines the temple tour with visits to the region's two other principal attractions, Champassak itself and another temple ruin, **Huei Thamo**. The site of this late ninth-century ruin lies in dense secondary forest close to a small riverside village about two and a half hours by boat beyond Champassak. It's known that the temple dates back to around AD 889, and was dedicated to Durga, consort of Shiva in her terrible aspect. But time and the elements, not to mention looters, have wreaked a violent toll, all but destroying the temple complex except for the broken walls and traces of three towers of one of the halls. Other relics — a lintel decorated with an image of Indra, and a stone linga with four faces at its head — are scattered among the trees.

The standard tour of this region involves a boat trip to **Muang Thamo**, also known as Um Muang or Muang Tomo, a ruined Khmer temple constructed in about the same period as Wat Phu. The ruins include a broad causeway bordered by linga and two carved stone sanctuaries that retain some degree of their original integrity. The return trip takes you back to Champassak where a car or mini-

van is usually waiting to take you to Wat Phu. The tour takes a full day, and the return to Pakse is generally by road via a vehicular ferry that crosses the Mekong at Champassak. With these grand relics of the Khmer culture, the quiet charm of Champassak and the stunningly beautiful vistas of the Mekong River and forested hills along the way, it's a day you'll remember for a long time.

It is also possible to rent a boat in Pakse, either chartered or take a public boat from the boat jetty on Pakse's Se Don River, and forget all about buses. Make the journey yourself, staying overnight in Champassak at the comfortable **Sala Wat Phu** for US$25 to US$30 per night and settle in. The knowledgeable and English-speaking manager Yoi

can help to organize you, inform you and do a lot to make your stay worthwhile. There is no need to book except at festival time. From Champassak to Wat Phu a distance of a few kilometers, it is fun to take a local sidecar motorbike or take a taxi from the ferry stop.

SI PHAN DON — THE 4,000 ISLANDS

Tours to the Mekong islands, **Si Phan Don**, further south are also available from Pakse or it is possible on a do-it-yourself basis where highlights include seeing the rare, freshwater dolphins (dry season) that come upriver to spawn and several spectacular cascades near the Cambodian border. While making the river hardly navigable at this juncture, the cascades are extremely attractive. Accommodations in the form of simple guest houses are readily available especially on the biggest island of **Don Khong** in the main village of Muang Khong, a former French settlement. **Donekhong Hotel** ((31) 212 077 on Khong Island at 3 Kanghong Street in Ban Khong Hong offers reasonable rates.

THAT LO RESORT

About 88 km (60 miles) northeast of Pakse is the delightful That Lo Resort — a piece of natural wilderness on the edge of the Boloven Plateau. Small but comfortable chalets look out across a river and cascading waterfalls, a popular stopover on many tours of the south. That Lo is in the midst of a coffee-growing area and a hill tribe region, with Katu and Alak villages worth exploring. Alternatively, it is rather pleasant spending time relaxing by the river. That Lo Resort can be booked through **Sodetour** ((31) 212 122 in Pakse or through the main office in Vientiane at ((21) 213 478 FAX (21) 216 314. They can also organize transportation. Rooms cost from US$15 to US$30 depending on the standard of facilities.

WHERE TO STAY

The former royal Boun Oum Palace, now known as the **Champassak Palace Hotel**

OPPOSITE: Temple carving at the Wat Phu Khmer ruins near Pakse. ABOVE: Transport on the Mekong River near Pakse.

((31) 212 263 or (31) 212 779/80 FAX (31) 212 781 (or in Vientiane ((21) 215 635 FAX (21) 215 636) is one of the fanciest places in Pakse, located on Route 13 on the outskirts of town. Occupying a prominent position overlooking the Se Don River, it is grand and grandiose with facilities but without much atmosphere, nor a much-needed swimming pool. Rooms sell for US$25 to US$35 plus tax and service.

The **Souksamlane Hotel** (/FAX (856-31) 212 281, 14 Ban Wat Luang, right in the center of town, is an old standby recently usurped by the Champassak Palace Hotel. Rooms cost US$18 a night with balconies overlooking the street, and there is a friendly ground-floor restaurant which provides mainly Lao food along with a few Western dishes. For my money, the **Auberge Sala Champa**, Pakse, just a two-minute walk from the busy market is the place to stay. The air-conditioned rooms go for US$25 to US$35. Try one of the older, unrenovated rooms upstairs (number 5) for a slightly seedy journey to the past, or the newer garden rooms with modern conveniences. This converted colonial villa offers comfort, charm, an air of antiquity and a powerful sense of what life must have been like in old Indochina — the woven bamboo walls and ceiling of its ground-floor bar, the sculptured stonework of its patio, the vast wire-netted balcony restaurant upstairs (advance reservation are required to eat there — allow at least eight hours and bring a party), the geckos and the lazy ceiling fans. The newer single-story annex built around the main house adds to the charm. A spacious double room here will cost you US$25 to US$35 a night.

Two kilometers (just over a mile) to the east, the **Residence du Champa** ((31) 212 120 FAX (31) 212 765 offers good management and comfortable rooms with a pleasant dining room all in a lush garden setting. Rooms range from US$30 a night to US$50 for a suite. The restaurant serves Lao, French, and Chinese food, and the hotel rents boats and buses for trips out of town. Opposite the market is the rather unattractive, Stalinist architecture of the **Pakse Hotel**, a huge, cavernous, place with rooms that in a pinch offer air-conditioning and plenty of space at

ridiculously low rates (from US$5 per night) if in the unlikely occurrence that all the other hotels are full. A newly-built hotel opposite could be worth trying.

WHERE TO EAT

With a large Chinese population, Pakse has several, if not glamorous, at least serviceable, restaurants scattered around the town center. The **Champassak Palace Hotel** has good Lao, Chinese and Thai food and a fully stocked bar. The Residence du **Champa** has a cozy dining room and a bar, popular with expatriates staying in Pakse. Opposite the market, the **Sedone Restaurant** is relaxed and friendly with reasonable Lao and Chinese-style food.

Otherwise several Chinese restaurants look promising; the best thing to do is to choose one that looks busy.

HOW TO GET THERE

Lao Aviation operates daily flights to Pakse, and now that the new road is completed it is even possible to travel by road from Vientiane — a journey which takes about 14 hours. It is also certainly possible to depart by ferry and road to Ubon Ratchahani in Thailand from where there are trains to Bangkok.

Several travel companies in Pakse can organize your tour, from **Sodetour** ((31) 212 122 in a French villa close to the boat jetty pier on Thasalakham Road. **Champa**

Residence ((31) 212 120 FAX (31) 212 765 can also organize a tour or a car at reasonable prices for guests. The **National Tourism Authority** of Laos Champassak Office ((31) 212 021 on Thasalakham Road (at the other end of the ferry pier) can rent cars for around US$80 per day.

In Luang Prabang, the Mekong River at sunset.

Travelers' Tips

GETTING THERE

While travel to Indochina is relatively uncomplicated, travel within the three countries can still be an adventure, although it is getting progressively easier to move about. For the main part, independent travel is quite possible and in some cases preferable. It must be noted however that the quality of experiences in more remote regions will be improved with the benefit of a good guide, or as part of a tour where arrangements are

gional hub, such as Bangkok (the closest), Singapore, or Hong Kong, and then book on a local carrier. Bangkok is probably the most convenient hub, as flights can readily be arranged to the main cities of Indochina. Travel companies like Diethelm who have a long experience in Indochina are also based in Bangkok, and for those who want organized itineraries, they are still probably the best (see TOUR COMPANIES, below). For direct flights, Vietnam Airlines flies Hanoi/Paris and Hanoi/Los Angeles via Taiwan. From Saigon they fly to Los Angeles via Manila, to

made for you. For the two- or three-week traveler, and certainly for those on shorter visits, Laos can be a destination in its own right, and if combined with a visit to Angkor Wat it will be both satisfying and enjoyable. To try and cover Vietnam as well within this time span will reduce the whole experience to hard work. To cover Vietnam alone could easily take three weeks unless one wants to pick out the highlights. Long-haul passengers will probably appreciate a few days at the end of their tour at one of the regional hubs (either Singapore, Bangkok or Hong Kong) for shopping and relaxation, or a beachside sojourn in Phuket, or Bali.

The best way to Indochina is to get your travel agent to book you on a flight to a re-

Vienna, Paris, Amsterdam and Zurich direct, and from Berlin, to Saigon via Paris.

Regionally based carriers like Cathay Pacific, Singapore Airlines, Silk Air, Malaysian Airlines, Royal Air Camboge, Philippine Airlines, Thai International, Royal Brunei and Lao Aviation are among those that fly to Indochina.

Thai International has daily service from Bangkok to Saigon and Hanoi, and Cathay Pacific has daily flights from Hong Kong to Ho Chi Minh City and Hanoi. Both services are run in cooperation with Vietnam Airlines. Thai International also operates daily flights

OPPOSITE: Fish market on the Cai River in Vietnam's historic Hoi An. ABOVE: Buddhist monks at Thien Mu Pagoda in Hue.

from Bangkok to Vientiane and three flights a day to Phnom Penh, again in cooperation with Lao Aviation and Royal Air Camboge.

Within Indochina, the three national airlines operate daily services linking Hanoi, Saigon, Vientiane and Phnom Penh, along with long-haul services to Moscow.

VISAS

Things have changed since the days when only overseas consulates provided visas to Vietnam, Laos, and Cambodia. Vietnam now

gives out two-month tourist visas, to be arranged before entry from Vientiane, Phnom Penh, Bangkok, Hong Kong, Singapore, or your own country's Vietnam Embassy. Once inside Vietnam, you are free to travel where you want without restrictions.

Cambodia grants two week visas costing US$20 on arrival in Phnom Penh which are extendable to up to one month at no extra cost.

Visas for Laos are also much easier to arrange. You'll probably only get a 15-day visa on arrival, but if arranged at the Lao Embassy in Bangkok, Singapore, or elsewhere, it is possible to get a month, and for a better price. Visas can be extended for US$3 per day, or alternatively, you can simply pay the fine of US$5 per day when departing. Visa prices vary considerably, depending on where the visa is obtained. For example, a visa on arrival at the airport costs you US$50. In Singapore a two-week visa costs S$65 (around US$35) from the Lao Embassy. No special passes are needed to travel around the country, but it is required you check in

with the immigration office in each town, which is automatic when you fly in.

LAO BORDER CROSSINGS

While Vientiane is the official entry point for Laos, Bangkok Airlines have recently started flights from Chiang Mai to Luang Prabang. Lao border crossings are opening up around this landlocked country, which shares borders with five countries. There are legal land crossings between all the neighboring countries except Cambodia. The most popular crossing is across the Friendship Bridge to Vientiane from Nong Khai in Thailand, an overnight train ride from Bangkok. The other popular alternative is to enter from Chiang Khong in northern Thailand via Huay Xai in the north of Laos, then take the boat downstream to Pak Bang and on to Luang Prabang. The third Thai border crossing is from Ubon Ratchahani at Chong Mek, about an hour from Pakse in southern Laos. There is one legal border crossing with Myanmar, which is also the one with China, by road at Botan in Luang Nam Tha Province (a visa for Laos can be obtained in Kunming in China). Border crossings with Vietnam are at Lak Sao on Road No. 8 in Bolikhamsay Province, which leads to the town of Vinh in Vietnam, and at Dan Savanh in Savannakhet Province on Road No. 4, leading to Quang Tri, north of I Iue in Central Vietnam. It is wise to have a valid visa and to check current policy before your arrival.

CUSTOMS

Customs requirements for all three countries look intimidating at first glance — a typical struggle with socialist regulations and paperwork. But that's all it is — forms in triplicate.

TOUR COMPANIES

As the key gateway to Indochina, Bangkok has the head office of **Diethelm Travel(** (66-2) 255-9150 FAX (66-2) 256-0248, Kian Gwan Building II, 140 Wireless Road, Bangkok 10500, who specialize in Indochina. Although expensive, they have excellent knowledge of the three countries.

More informal tours are offered by Australian-based **Intrepid Tours** ((61 3) 9416 2655 FAX (61 3) 9419 4426 E-MAIL info@intrepid travel.com. While the British **Exodus Adventure** has offices all over Europe, in Australia and New Zealand, the head office is in London ((44 181) 675 5550 FAX (44 181) 673 0779, or for brochures telephone ((44 181) 673 0859 E-MAIL sales@ exodustravels.co.uk.

TOUR AGENCIES IN VIETNAM

Within Vietnam, travel companies are government operated. Each province has at least one. However some independent tour agencies are now operating in the country, as joint ventures with the government.

In Hanoi the government-owned **Hanoi Tourism** ((4) 826 6714 FAX (4) 825 4209 can be found at 18 Ly Thuong Kiet. **Exotissimo** has an office in Hanoi ((4) 828 2150 FAX (4) 928 0056 at 26 Tran Nuat Duat Street and in Saigon ((8) 825 1723 FAX (8) 829 5800 at 2 bis Dinh Tien Hoang Street, District 1. The well-run and recommended **Especen Tourist** ((4) 826 6856 FAX (4) 826 9612 is at 79 Hang Trung, Hanoi. **Buffalo Tours** ((4) 828 0702 FAX (4) 826 9370 is at 11 Hang Muoi, Hanoi.

In Saigon, the aggressive and expansive **Saigon Tourist** ((8) 829 8129 FAX (8) 822 4987 is at 49 Le Thanh Tan. This giant government agency owns hotels all over Vietnam and has a wide selection of specialty and tailor-made tours to suit all tastes. **Global Holidays** ((8) 822 8453 FAX (8) 822 8545, 106 Nguyen Hue Street, District 1, Saigon, is an Australian-owned and managed concern.

TOUR AGENCIES IN LAOS

Diethelm Travel ((21) 213 833 or (21) 217 151 FAX (21) 216 294, PO Box 2657, on Setthathirat Road, Nam Phu Circle, in Vientiane, is not only the most expensive, but also the most forward-looking. I met the Bangkok head honcho in Sekong in the far south of Laos while he was scouting for new tour options. **Inter Laos Tourism** ((21) 214 832 or (21) 214 232 FAX (21) 216 306 or (21) 214 232, at Setthathirat Road (Nam Phu Circle), is also helpful and seems to be well run. **Sodetour** ((21) 216 314 or (21) 213 478 FAX (21) 216 313 or (21) 215 123, started by a

Frenchman, has the head office in Vientiane at 114 Quai Fa Ngum. They have other offices in Pakse and Luang Prabang, a hotel in Xieng Khuang and another known as That Lo Resort in the south.

Lane Xang Travel ((21) 212 469 or (21) 213 198 FAX (21) 215 804 or (21) 214 509 at Pang Kham Road also offers a wide range of tours.

TOUR AGENCIES IN CAMBODIA

Cambodia's domestic tour companies act as ground-handling agents for international

operators and also arrange independent itineraries — mainly to Angkor Wat.

Apsara Tours ((23) 426 562 or (23) 722 019 FAX (23) 426 705 or (23) 427 835 can be found at Street 8, RV Vinnavaut Oum, Phnom Penh. The head office of **Angkor Tourism** (/FAX (63) 380 027 is in Siem Reap at Street 6, Phum Sala Kanseng. The Phnom Penh office (/FAX (23) 362 169 or (23) 427 676 is at 178c Vithei Trasak Phaem.

Diethelm Travel (Cambodia) ((23) 426 648 FAX (23) 426 676, 8 Samdech Sothearos Boulevard, Phnom Penh, offers excellent service for a price. **J&C Paradise Travel** ((23)

OPPOSITE: Schoolgirls in traditional *ao dais* in central Saigon. ABOVE: Village girl distills rice wine in Luang Prabang region.

722 302 or (23) 722 402 FAX (23) 722 702 is located at 13 EO Street Sangkat, Phsar Thmei II, opposite Dusit Hotel.

ACCOMMODATIONS

As this book went press, there was still no specific reservation information on the new **Hanoi Hilton**, however listed below are Hilton's worldwide reservation numbers:

Australia (1800 22 2255;
England (0990 445 866;
France (0800 90 75 46;
Singapore (1800 737 1818;
Switzerland (0800 55 2246.
United States ((800) HILTONS.

Below you will find the contact numbers for **Resinter Worldwide Reservations**.

Australia (1800 64 22 44 FAX 1800 15 07 07;
Austria (06 60 85 94;
Belgium (08 00 101 27 (French); 08 00 195 11 (Dutch);
Denmark (800 185 95;
Finland (9800 144 04;
France (01 60 77 87 65 FAX 01 69 91 05 63;
Germany ((6196) 460 36 FAX (6196) 483 817;
Great Britain ((181) 741 9699 FAX (181) 748 9116;
Hong Kong (800 25 09 FAX 800 46 33;
Indonesia (001 800 618 57 FAX 001 800 618 67;
Italy ((2) 29 51 22 80 FAX (2) 29 51 08 57;
Japan (003 161 6353 FAX 003 161 6354;
Malaysia (800 25 78 FAX 800 51 41;
Netherlands ((20) 549 54 49 FAX (20) 642 83 75;
Singapore (800 61 61 367 FAX 800 61 61 202;
Spain (900 67 67 67 FAX 84-1 326 51 63;
Sweden (020 793 153;
Switzerland (155 64 77 (French); 155 80 22 (German);
United States ((800) 221-4542 FAX (914) 472-0451

GETTING AROUND

The national carriers—Vietnam Airlines, Lao Aviation, and Royal Air Camboge — operate very efficient domestic networks in Indochina. They can also be quite flexible. Lao Aviation and Royal Air Camboge add extra flights to their key cultural destinations,

Luang Prabang, and Angkor Wat, if the volume of tourists gets too big for scheduled services.

Their recently-upgraded aircraft are in reasonable condition. For all that, you get a good aerial bus ride to where you want to go, and despite reports to the contrary, the safety record in all three countries is high when you consider the volume of passenger flights. Modernization of all three networks is just around the corner. Domestic airports right across the region are rundown and rustic, giving a delicious feeling of being there before the crowds. They're quite comfortable, and food and drink are generally available. While I was waiting for a (delayed) plane at Oudomxay Airport, gazing out at the bright new runway, I was surprised by the plane landing behind me on the old grass airstrip. Even though the runway was ready for use, it hadn't been "handed over" to the relevant authorities, so the bumpy grass patch complete with goats and grass cutters was still very much in use!

Airline booking offices in main cities are as follows:

Vietnam Airlines Head Office is in Hanoi ((4) 825 0888 FAX (4) 824 8989 at 1 Quang Trung Street; the Saigon office ((8) 823 0696 FAX (8) 829 2118 is at 116 Nguyen Hue Street (opposite the Rex Hotel). Vietnam Airlines is well run with a network of flights to many destinations across the country including Phu Quoc Island.

Lao Aviation ((21) 212 051 FAX (21) 212 056 is at 2 Pang Kham Road, Vientiane. Lao Aviation has a good network of flights across the country which will get most visitors close enough to where they want to go. The company is not yet computerized and once made, bookings tend to stand firm. One person I spoke to faxed a booking from Malaysia to Vientiane for an internal flight which was honored with no problem when she went to the office in Vientiane.

Royal Air Camboge ((23) 428 891 FAX (23) 428 895, at 62 Tou Samouth Boulevard, Phnom Penh, is a well run regional airline.

To travel between Saigon and Hanoi, you can take the Reunification Train or brave the arduous public bus system.

By Road

Rental cars, minibuses, and limousines are available throughout the region, with the main hotels and tour companies operating them. If you're traveling independently, charges seem to be the same whether you're in Hanoi, Vientiane or Phnom Penh — around US$50 to US$80 a day, depending on whether it's a minibus or a new Toyota. This includes gas and the driver's accommodation if there's an overnight stay.

Private taxis are available in most cities, on-call around the main hotels, but these generally involve a lot of haggling over the fare. And why cram yourself into a taxicab when you can roll along in the open air aboard a pedal-cyclo, which you'll find just about everywhere you go in Indochina, and will cost you about US$1 per hour ($10 a day). Otherwise, it's easy to rent motorcycles in Vientiane and Phnom Penh at about US$5 to US$10 per day (24 hours), and most cities rent bicycles if you want to get around a little more sedately.

SECURITY

Indochina requires the sort of precautions you'd take against theft anywhere else in the world — put anything of particular value in the hotel safe deposit, don't carry a lot of money on you in the streets, and keep your camera close to you at all times. Laos has a very low crime rate and, ironically, so have Phnom Penh and Angkor Wat, except for the underpaid and probably very hungry soldiers, but Saigon requires extra care day and night — I have had five pairs of glasses and sunglasses, pens, and camera lenses stolen, as well as an attempted bag snatch by motorbike, in Saigon alone, and that is without being particularly careless. Resident expatriates seem to accept it as part of the price they pay for living there. Thieves are extremely accomplished and very fast, it is all part of the Vietnamese survival skills. Be warned and come prepared. Theft from hotel rooms is not unknown, particularly in Saigon's older, more down-market hotels, and there have been several quite serious hotel thefts in Da Nang.

In Phnom Penh, visitors are warned not to leave their hotels at night, and to travel in groups and stay away from dark streets. If you are going out take a car or *moto* driver you can trust — preferably one recommended by the hotel.

Laos has very few security problems although take a guide if you'll be walking in remote areas that may be mined.

HEALTH

Although the standards of hygiene throughout Indochina are low compared to Western countries, international standard medical services for tourists are available in all the capital cities A few sensible and basic precautions should help to guard against any problem bigger than an occasional case of upset stomach. A reputably effective guard against a lot of stomach bugs is to eat lots of chili, which acts as a mild antiseptic. Hepatitis is the most prevalent disease to guard against, and no one should travel the region without prior vaccination. Although you'll find mosquito nets in most older hotels, malaria and encephalitis are common enough that you'll need encephalitis shots before traveling and a course of malaria pills while you're on the road. At least take a good repellant and keep it handy. The cooler dry season in the northern winter months has noticeably fewer mosquitoes. I have traveled several times without contacting anything awful, but it is important to cover up against mosquitoes at dawn and dusk, and if mosquitoes are about, get out that repellant at once — it doesn't help while it sits in your bag.

To put all this into perspective, these are the sort of health precautions that one would expect to take in any developing region. With prudence, the worst thing you'll have to safeguard yourself from is an upset stomach and accompanying diarrhea, and the best way to do that is to drink the readily available bottled water wherever you go. It is probably wise to avoid ice, but I used it every day without a problem. An old Asian trick I learned in Singapore is to let the ice melt a little then drain it before pouring your drink into it. I was also quite wary about fruit juices in market places. If buying food in out-of-the-way places, buy food that is newly-fried or barbecued to ensure germs are at a minimum, rather than food that has been standing for some time. Common sense and a keen sense of survival go a long way.

MEDICAL MATTERS

Most international hotels have an in-house clinic, or at least a recommended doctor on call for emergencies; otherwise you can take out travel insurance with the Hong Kong- and Singapore-based Asia Emergency Assistance, which has a joint venture in Vietnam providing, among other things, medical evacuation in event of serious illness or injury.

Hanoi and Saigon
Asia Emergency Assistance has four international clinics in Vietnam open 24 hours a day, with fully-equipped emergency wards, international standards of medical and dental care, and emergency assistance and evacuation: **Hanoi Clinic** ((4) 934 0555 FAX (4) 934 0556 at 31 Hai Ba Trung, Hoan Kiem District; **Saigon Clinic** ((8) 829 8520 FAX (8) 829 8551, 65 Nguyen Du Street. The **Vietnam International Hospital** ((4) 574 0740 FAX (4) 869 8443 in Hanoi is at Phong Mai Road, Dong Ha; and in Saigon, the **Cho Ray Hospital** ((8) 855 4137/8 FAX (8) 855 7267 at 210B Nguyen Chi Thanh Boulevard has a clinic for foreigners with English-speaking doctors.

Vientiane
Vientiane has two emergency clinics available for foreigners. The **Australian Clinic** ((21) 413 603 is attached to the Australian Embassy Compound and the **Swedish**

Clinic ((21) 315 015 is located near the Swedish Embassy.

Phnom Penh
In Phnom Penh, the **Tropical Traveler's Medical Clinic** (/FAX (23) 366 802, at 88 Street 108, has well qualified doctors including specialists in tropical medicine.

TOILETS

In Indochina toilets vary between the conventional Western kind in hotels and places

in the cities to the simple squatter ("Turkish") type they tend to be in rural areas. Cleanliness is generally of an acceptable standard and not too traumatic.

CURRENCY

In a region in which the currencies are called the *dong, kip* and *riel*, it's comforting (or worrying) to know that the United States greenback will get you just about anything you want. The reason is that all three currencies are worthless anywhere else. In 1998, the Vietnamese *dong* was running at around 12,000 to the United States dollar, and the Lao *kip* at about 2,000, and in Phnom Penh a dollar would get you 2,000 or more *riels*. Laos and Cambodia were also annoyingly quoting prices in Thai baht as an alternative to the greenback.

In Vietnam, the newish 50,000 *dong* note makes it considerably easier to carry local

OPPOSITE: Plowing with water buffalo in ricelands near Nha Trang. ABOVE: Farming near Da Nang.

currency without feeling as though you've robbed a bank (or that someone will rob you). In Vientiane and Phnom Penh you can pay for a meal in a restaurant with United States dollars and request your change in a mixture of dollars, Thai baht and the local currency — and get it all at the exact exchange rates.

As for currency exchange, most local and foreign joint-venture banks in Vietnam, along with the better-class hotels, will convert the main international currencies at a cost — it is always better to stick to money changers

If your hotel has no communication facilities, then upmarket hotels will let you use their Business Center, quite often for a very reasonable fee. E-mail centers are available all over Hanoi and, to a lesser degree, in Saigon.

Vietnam is well covered by news media, most of which are based in Hanoi. For foreign language papers, i.e. English, there are two run by foreign journalists, with a high degree of input by local journalists. The bi-weekly *Vietnam Economic Times* is probably the most responsible reporting with the older

when possible; most banks take forever to change.

Virtually all major hotels and restaurants in Laos and Cambodia accept credit cards and can often cash traveler's checks as well.

COMMUNICATIONS AND MEDIA

International dial-direct telephones and fax services are now fairly common in the up-market hotels in Vietnam and Cambodia, though an incoming or outgoing fax will often take up to 24 hours to get through in Vietnam. This is generally the time it takes for scrutinizers in the Post Office to translate it before passing it on. However, there's no such petty censorship in Phnom Penh.

Vietnam Investment Review also giving excellent coverage, with a slant, not surprisingly, towards investment. Both are worth grabbing on arrival in Vietnam too see what is current. They also both run excellent monthly entertainment guides that come with the paper. Saigon's best magazine stall is opposite the Rex Hotel where copies of current papers and magazines gleaned from incoming planes are sold as well as current *Newsweek* and *Time* magazines. You can dig through the piles on display and find marvelous back issues with articles that you missed the first time round.

Phnom Penh, likewise, has international telephone and fax services, with most hotels hooked up; but rates are as extortionate as

Vietnam. Media in Phnom Penh consists of the biweekly *Phnom Penh Post*, started during the days of UNTAC by a group of intrepid journalists, and the newer tourist/news rag, the *Bayon Pearnik*, which seems designed to entertain the staff as much as the reader. The best bookshop in town is located at the Cambodiana Hotel with a marvelous array of books on Cambodia and international news magazines.

Telecommunication is remarkably cheap in Laos, compared to the extortionate rates charged in Vietnam and Cambodia (US$6

220 volts. If you're taking a hairdryer, electric shaver, or camcorder battery charger with you, make sure it's will work with either 110 or 220 volts.

Laos and Cambodia are a lot simpler — it's 220 volts wherever you go, though the antiquated wiring may have you wondering about overloads and short-circuits. In Phnom Penh, particularly, blackouts occur quite regularly, and most older hotels provide candles in the rooms. Of course, it also means no air-conditioning when the power fails.

per minute for international calls). E-mail facilities should be well established in Laos by now, otherwise it is a matter of going through Bangkok, as many of the residents were doing when I was there. News publications available in Laos include the *Bangkok Post*. *Time* and *Newsweek* are available at hotel bookstores and Raintree (in Laos Plaza Hotel Lobby) has a decent array of books.

ELECTRICITY

Vietnam has two currents, 110 volts and 220 volts, depending what city you're in, with two-pin round-prong outlets for 110 volts and two-pin United States-type outlets for

TIME

All three countries are on Greenwich Mean Time + 7 hours.

PHOTOGRAPHY

You'll find plenty of shops and hotel boutiques in every city and main town of the region selling Kodak and Fuji print film, along with specialist camera stores with up-to-date cameras and equipment. However, Kodak and Fuji slide film isn't always easy to come by, and when buying film of any sort check the expiry date and storage

OPPOSITE: Century Hotel and typical river life in Hue. ABOVE: Lan Xang Hotel pool in Vientiane.

conditions — film that has been lying around too long in the heat and humidity is as good as useless since it tends to lose its color — you can easily get palmed off with film that should have been trashed a couple of years ago.

Most key airports now proudly label their X-ray machines "Film Safe," but if I were you I wouldn't trust them. Hand your film over in a polyethylene bag for security inspection.

TRADITIONS AND TABOOS

The Indochinese are a very hospitable people, and quite flexible in their dealings with foreigners. This means that when problems or frustrations occur — and they will, just as the first travelers into mainland China found — you should try to treat the issue as problem-solving, not the flashpoint for anger. There's an innate and quite compatible sense of humor in this region which you won't find in eastern and northeastern Asia, and in most cases it smooths out misunderstandings before they reach boiling point. A bit of patience and tolerance goes a long way. So does a smile.

When Buddhist temple touring in each country, please remember that it's a mark of respect to dress for the occasion — shirts and slacks (or jeans) rather than shorts and tank-tops, and women should cover their arms and legs. Monks and novices are forbidden to accept anything directly, hand-to-hand, from women. In Laos and Cambodia, it's very disrespectful to sit with your feet pointing at other people (be careful too when crossing your legs), and the local people will naturally contort themselves (when sitting on a boat or wherever) so as not to point their foot.

It's also bad form to touch or pat people on the head, the head is a holy place and even a friendly pat on the head for a child is not to be entertained.

WOMEN ALONE

Women traveling alone should find few difficulties in predominantly Buddhist Indochina. If one follows basic standards of dress — modest clothes, like a dress or long shorts or trousers rather than short shorts and hal-

ter tops, and behaves in a respectable and respectful fashion there is no reason why the trip should not prove to be delightful. One good ploy to adopt for single women is an imaginary husband and a child or two; as most Asians find it almost inconceivable that a woman could be traveling alone, and to be without a family is to be a subject of pity. Better to invent a big happy family and retain the status quo rather than rush headlong into a major cultural collision or try to reeducate the whole of Asia when their more traditional ways have a lot going for them.

WHEN TO GO

Although Indochina lies in the subtropical zone, the climate has distinct variations and when you go depends largely on what weather you prefer. The northern areas of Vietnam and Laos catch the tail end of the northern monsoons from China and Central Asia, which in the winter months bring cold and wet weather and temperatures that drop as low as 8°C (46°F) to 15°C (59°F). It can get particularly cold in the Lao mountains, in Muang Sing and on the barren Xieng Khuang Plateau and the Plain of Jars, where an icy wind cuts through to your bones. Northern Vietnam, in the mountain areas of Sa Pa and Dien Bien Phu, is similarly chilly. Hanoi and

Luang Prabang too can be cold and gray in the winter but after a week, the sun can shine through and the weather becomes unseasonably warm.

For the most part, there are really only two seasons to Indochina: hot and wet and hot and dry. In the summer, Hanoi gets extremely hot and humid, with temperatures often above 30°C (86°F), and the heat and humidity intensify as you head south. The southern monsoon brings a great deal of rain, but it's a fairly benign season. In Vietnam, Laos and Cambodia you generally get one

cyclos — is something that should be kept in mind at all times.

WHAT TO BRING

Light cotton tropical clothing is most suitable for southern travel throughout Indochina, though if you're traveling north in the winter months prepare for cold conditions. Dress in layers that can be added and removed with the temperature fluctuations. A warm jacket, socks, a few knits and a hat do not go amiss. Even in the summer months

refreshing cloudburst a day, usually in the late afternoon, with fine weather either side of it. However, Vietnam lies at the end of the path of the Asia-Pacific typhoons, which often dash themselves along its central and northern coasts in the summer months bringing powerful gales, torrential rain and flooding.

The dry season, from October to early April, is the best time to travel through the region. The days are warm to hot and the nights cool, though you can get an occasional day of chilly fog and rain. As an added bonus, the countryside turns lush and green after the raining season. Whatever the season, heat and the risk of dehydration — especially if you travel around on motorbikes or open

a pullover or light jacket should be carried if you're going to destinations such as Da Lat in Vietnam's Central Highlands or Sa Pa in the north. The mountains of Laos too, can get chilly at night. I found that a thick jacket and hat were very welcome in February, although Vientiane was very warm. It's advisable to dress quite modestly — no revealing tank-tops or crotch-hugging shorts or hotpants. The Indochinese get a kick out of fashionable foreigners, but do not appreciate semi-nudity.

OPPOSITE: Monk in woolly hat in Luang Prabang's early morning chill. ABOVE: The tranquility of Indochina — the Mekong in Luang Prabang.

LANGUAGE BASICS

The languages of Indochina are not only quite difficult to learn, they are also tonal, with barely perceptible differences to the Western ear, yet with major differences to those attuned. To make matters even more interesting, the languages are each of different derivations and share few if any similarities. Even learning to count up to ten in each language can become a little confusing. That said, I will endeavor to give a few words that can make a difference. The spelling is phonetic.

In Laos the most important word to learn is *sabadee*, which is a gracious and all-purpose greeting that shows respect and friendliness. Unlike addresses in Vietnam, it can be used with all ages and classes of people and almost guarantees a pleasant response. The other important phrase is *"kopp chai lai lai"*— or "thank you" — also guaranteed to bring a smile to the face of the recipient.

The ubiquitous noodle soup translates as *pho*, pronounced "fur," while chicken is *kai* and the spicy Lao salad is pronounced *laarp*. To count to ten: 1—*neung*, 2—*sorng*, 3—*saam*, 4—*sii*, 5—*haa*, 6—*hok*, 7—*chet*, 8—*pet*, 9—*kao*, 10—*sip*.

In Vietnam, with their system of honorifics and titles, there is no all-purpose greeting except for *"ciao"* — which no one really seems to use much, and it certainly lacks the universal grace of "sabadee." With the extremely complex tonal system of Vietnamese language, one word like *ga* can mean "chicken" and "railway station" (from the French, *gare*) and the almost indistinguishable *ca* means fish, while *cua* means crab. It's all in the pronunciation. Rice is *com* and the noodle soups are known as *pho*. Beef is *bo* and white rice noodles are *bun*. Spring rolls are *cha gio*, pronounced as "char yio." Numbers from one to ten are: 1—*mot*, 2—*hai*, 3—*ba*, 4—*bon*, 5—*nam*, 6—*sau*, 7—*bay*, 8—*tam*, 9—*chin*, 10—*muoi* or *chuc*. If you can master all these, there are some excellent phrase books available for the dedicated.

Khmer is different again and a quick visit to Angkor Wat barely warrants the bother. However, the numbers from one to ten are: 1—*mouy*, 2—*pee*, 3—*bay*, 4—*boun*, 5—*bram*, 6—*bram-mouy*, 7—*bram-pee*, 8—*bram-bei*, 9—*bram-boun*, 10—*duop*. Chicken is *maan*, fish is *trey*, noodles are *mee* or *moum banjook*, and a pagoda or monastery is called a *wat*. Good luck.

Recommended Reading

BAO NINH. *The Sorrow of War*. London: Secker and Warburg, 1993. The Vietnam war from a Vietnamese perspective.

CONNORS, MARY F. *Lao Textiles and Traditions*. Oxford: Oxford University Press, 1996.

Ethnic Minorities in Vietnam. Hanoi: Foreign Languages Publishing House, 1984.

GREENE, GRAHAM. *The Quiet American*. 1954.

HALL, D. G. E. 1955. *A History of Southeast Asia*. New York: Saint Martin's Press, 1981.

HERR, MICHAEL. *Dispatches*. Picador, 1968.

KREMMER, CHRISTOPHER. *Stalking the Elephant Kings: In Search of Laos*. Allen and Unwin Publishing, 1977.

MAXWELL, ROBIN. *Textiles of Southeast Asia*. Oxford: Oxford University Press.

OSBORNE, MILTON. *River Road to China: The Search for the Source of the Mekong 1866-73*. Allan and Unwin Publishing, 1975. An account of this nineteenth century exploration.

SHEEHAN, NEIL. *A Bright and Shining Lie*. New York: Random House, 1988.

THEROUX, PAUL. *The Great Railway Bazaar*. Penguin Books, 1975. Includes an account of a his ride on the Reunification railway before reunification.

WADE, JUSTIN. *The Vietnam Wars*. London: Weidenfield and Nicolson, 1991.

Quick Reference A–Z Guide
to Places and Topics of Interest with Listed Accommodation, Restaurants and Useful Telephone Numbers

Photography Credits

All photographs are by **Alain Evrard** with the exception of the following.

Jill Gocher: 10, 11, 12 *top and bottom*, 13, 14, 16, 17, 19, 20, 21, 22, 23, 24 *top and bottom*, 25, 26, 27, 28, 29, 31, 32, 33, 34 *top and bottom*, 35, 37, 38, 39, 41, 42 *top and bottom*, 43, 44, 45, 46, 47, 48, 50 *top and bottom*, 51, 52, 54, 55, 57, 58, 59, 60, 61, 62, 63, 65, 66, 67, 68, 69, 75, 76, 138, 140, 145, 163, 181, 187 *top*, 195, 206, 212 *right*, 218, 263.

Nick Wheeler: cover, 7 *left and right*, 108, 109, 112-113, 118 *left and right*, 127, 130, 141, 142–143, 146–147, 148, 157, 178, 211, 224–225, 225, 226, 227, 228, 229, 230–231, 233, 234 *left and right*, 235, 236–237, 238 *left and right*, 239, 242–243, 245, 246 *left and right*, 248, 249 *left and right*, 250–251, 252, 253, 254–255, 257 *top and bottom*, 259, 260, 261, 264, 265, 275, 276, 277.

Tim Hall: 72.